Merry Christmas Chet

Antique Farm Equipment
The Elsbree Collection

❧

Chuck Elsbree

Rayve Productions
Windsor, California

About the Cover

Once upon a time, she was a proud part of the farm team that provided food and income for the family, but when she grew old and new machines replaced her, she was abandoned on a shady hillside where vital parts were wrenched from her and, as days melted into years, she bled rust into the soil. Eventually she was rediscovered, and newly appreciated for the contributions she had made. Now she invites younger generations to embrace her — just as she is with all her flaws — to learn about the work she did and how she and her farm teammates helped make America great.

Rayve Productions Inc.
Box 726, Windsor, CA 95492

Copyright © 2012 Chuck Elsbree

Book Design: Mary McEwen — Design Girl Graphics
Elsbree Farm Equipment Photographs: Susan Nelson

All rights reserved. This book is protected by copyright. No part of this publication may be reproduced or transmitted in any form or by any means, electronic or mechanical, including photocopy, recording, or any information storage and retrieval system now known or to be invented, without written permission from the publisher, except by a reviewer who wishes to quote brief passages in connection with a review written for a magazine, newspaper, broadcast, or the Internet.

Library of Congress Cataloging-in-Publication Data
Elsbree, Chuck, 1937-
Antique farm equipment : the Elsbree collection / by Chuck Elsbree.
p.   cm.
Includes bibliographical references and index.
ISBN 978-1-877810-41-1 (hardcover : alk. paper)
1. Farm equipment--Collectibles--United States. 2. Antique and classic tractors--Collectors and collecting--United States. 3. Agricultural machinery--Collectors and collecting--United States.
4. Elsbree, Chuck, 1937-
I. Title

S751.E57 2012
631.3--dc23
Library of Congress Control Number: 2011013118

BISAC Subject Headings: ANT000000, ANT001000, ANT008000

Printed in Hong Kong

About the Author

Chuck Elsbree grew up working on the family ranch in Windsor, California during the 1940s and 1950s. Work on the farm and in the surrounding area involved the use of many of the implements in his antique farm equipment collection, which had been modified for tractor use.

For most of his career, Chuck worked as a research and development engineer in the aerospace industry. In this capacity Chuck helped develop the missiles used in the Apollo Space Flight Program. His last aerospace assignment was as program manager for developing the hardware used on the Lunar Lander. Chuck later started his own electronic hardware company, and has subsequently been granted six different patents.

Chuck is once again ranching in Windsor, using his research and development skills to raise premium wine grapes and produce quality wines.

Elsbree Family Snapshots

On the ranch with Brutus

Salud!

Chuck, Aerospace Engineer

Diana and Brutus

Antiquing in New Mexico

Dedication

To my wife, Diana . . . without whose help and support this book would not have been possible. She typed the original manuscript and was my partner on numerous excursions looking for specific old farm equipment, enduring many nights of not-so-first-class meals and accommodations in remote parts of the country.

To Sue Nelson, for her professional photographs and her assistance with early editing.

To my brothers Don and Jack . . . for their help with farm equipment identification, and son Mike, for his talented restorations on most of the equipment.

Introduction

For the most part, the collection presented here includes turn-of-the-century farm, ranch and other related antique implements manufactured prior to early tractor development. All implements shown (and more) are on display at the Elsbree Family Vineyard Ranch in Windsor, California.

The collection took over thirty years to assemble. Many people considered these items junk and, as such, they had been abandoned and allowed to deteriorate significantly. As a result, it has become increasingly difficult to find pieces in any reasonable condition. While many items came from local farms and ranches, most were obtained from outside the immediate region.

By preserving the implements and describing their functional and mechanical purposes, I hope that readers, and those who tour the collection on site, will better understand and appreciate the engineering talent required to create and manufacture antique farm equipment.

I also hope that current and future generations will recognize the great contributions of these engineering marvels in helping build a strong America. Those early farm tools were the prototypes of most modern implements in use today. Look closely and you will see!

General Information About Horse-Powered Implements

Most implements with handles were pulled by one or two horses. Where there was a seat for a rider or a single or dual functioning element attached, the unit was pulled by two or three horses. Gang cultivators could use up to eight horses. Almost all horse-drawn implements had height adjustments for single trees to allow for the height of the horse and thereby control implement depth and angle.

Equipment such as hay balers, hay blowers, and sprayers were moved to stationary positions for their particular functions and could be powered by four to six horses.

After the development of engine-powered tractors, almost all horse-powered equipment eventually had the tongues modified so that they could continue to be used with tractors.

A Word About the Accuracy of Identification

The implements on site show the wear and tear that occurred over a century of time, with some items in working condition and others missing a small part or two. Despite this, they all continue to show their intended use and function. Great effort has been taken to preserve the original profile of each item.

Because most antique farm implements do not have the date, name, or manufacturer stamped on them for identification, the information presented in this collection reflects the author's best-effort research. Many manufacturers copied features of other manufacturers, which makes it difficult to be fully accurate in determining names and dates for certain items. Patents were hard to enforce and many manufacturers did not bother to obtain protection. Some of the dates listed reflect the manufacturers' known production period of similar equipment.

Contents

Seeders
- Pratt's Seed Sower — 6
- Single Row Seeder — 7
- Single Seeder — 8
- Seed Planter — 9
- Iron Ace — 10
- Avery Sure Drop Planter — 11
- Grain or Seed Drill — 12
- Keystone Corn Planter — 13
- Check Row Marker — 14

Cultivators
- Dual Purpose Cultivator — 15
- No. 101 Planet Junior — 16
- Garden Cultivators, Case-Keeler & Henderson — 17
- Planter's Pride — 18
- Surface Cultivator; Double Fork Rake — 19
- Jack Rabbit Cultivator — 20
- Buckeye Cultivator — 21
- Small Cultivator — 22
- Complex Cultivator — 23
- Belanger Steel Scuffler; Cultivator — 24
- No. 101 Planet Junior Horse Hoe — 25

Harrows
- Spring Tooth Harrow — 26
- Revolving Harrow; A-Frame Harrow — 27
- Ten Blade Disc — 28
- Pennsylvania Disc Harrow — 29
- Harrow Sulky — 30
- Iron Age Harrow — 31
- Harrow Cultivator; Spike Tooth Harrow — 32

Plows
- Prairie Breaker Plow — 33
- Wooden Double Shovel Plow; Walking Furrow Plow — 34
- Canton Clipper Plow; Reversible Plow — 35
- Red Jacket Walking Plow; Furrow Plow — 36
- Stirring Plow; Double Mold Plow — 37
- Ditch Irrigation Plow; Ditch Cleaner Plow — 38
- Potato Digger Walking Plow; Potato Digger Plow — 39
- Large Walking Plow; Turning Plow — 40
- Improved Winged Shovel & Potato Digger Plow; Vulcan Chilled Plow — 41
- Oliver Sulky Plow — 42
- Cambridge Reversible Plow — 43
- Tractor Gang Plow — 44

Haying Equipment
- California Grapple Fork — 45
- Hay Rake/Sling — 46
- Hay Blower — 47
- Sandwich Hay Mower — 48
- Ideal Mower — 49
- Hay Tedder — 50
- Pull Power Hay Press — 51
- King of the Meadow; Hand Dump Rake — 52

Other Equipment
- Subsoil Ripper — 53
- Wood Tank Sprayer — 54
- Corn Kernel & Husk Remover; Stalk Cutter — 55
- Windmill & Pump — 56
- Fertilizer Spreader - Model 11 — 57
- Bemis Plant Setter — 58
- Fordson Tractor — 59
- Ox-Head Yoke; Single Trees — 60
- Grader — 61
- Slip Scraper; Manual Fresno; Fresno Canal — 62
- Road Leveler — 63
- Flatbed Wagon; Steel Carriage Jack — 64
- Cargo Wagon — 65
- Hand-Drawn Fire Hose Cart — 66
- Trapping Implements; Copper Stills — 67
- Gold Mining Bucket; Bucket Winch; Markers; Wine Press; European Wine Press; Scythe & Stable Fork — 68
- Forge; Timber Saws; Ice Cutter & Ice Hook; Automatic Clothes Washer & Boiler; Wheel Barrow Sprayer — 69

Fordson Tractor, c. 1925.

Seed Planters

Pratt's Seed Sower
Leoa Pratt
◁1844▷

The one-horse drill and check planter is one of the earliest recorded seed planters. The wood box contained seed and dropped it into the drill. Then, activated by a chain driver, the drill pushed the seed into the ground.

A. Spindle seed push down
B. Side harrow gathers dirt
C. Seeder box rotor drop controls

Single Row Seeder

Peter Henderson & Co. 1847

On this implement, the front runner opened the furrow and the drill dropped seed, then pushed soil over it. The curved rear wheel pressed the soil over the seed. The seed-drop spacing control was preset and driven by a revolving rear-wheel toggle. The rear wheel was kept clean with a mounted scraper fitting on the curved wheel.

A. Runner
B. Drill
C. Seed canister controls
D. Curved wheel & scraper
E. Hookup control for depth

Single Seeder
Case-Keeler
1847

The Single Seeder had a chain sprocket on the front wheel that spaced seed drop after the furrow was opened with the seed-drop drill. The side harrow covered seed, and the curved rear wheel compressed dirt over the seed.

A. Chain drive
B. Trench digger
C. Seed adjustment
D. Dirt spreader
E. Packing wheel

Seed Planter
Oliver Chilled Plow Works
◁1917▷

The Seed Planter was pulled with a single horse and used a front disc to open a furrow for the seed drop. One canister was used for seeds and the other for fertilizer and drops were adjusted by various-distance sprockets on wheel openings at the bottom of the canisters.

A. Rotary settings for seed drop
B. Seed drop and furrow blades
C. Seed and fertilizer canisters

Iron Ace
Case-Keeler
1847▷

The manually operated Iron Ace had a chain wheel that controlled seed drop frequency. The drill in front of the seed drop loosened the soil and made a furrow. Then, the curved rear wheel covered the seed with soil. This Iron Ace is missing its chain drive.

A. Drill and seed drop
B. Rear curved wheel

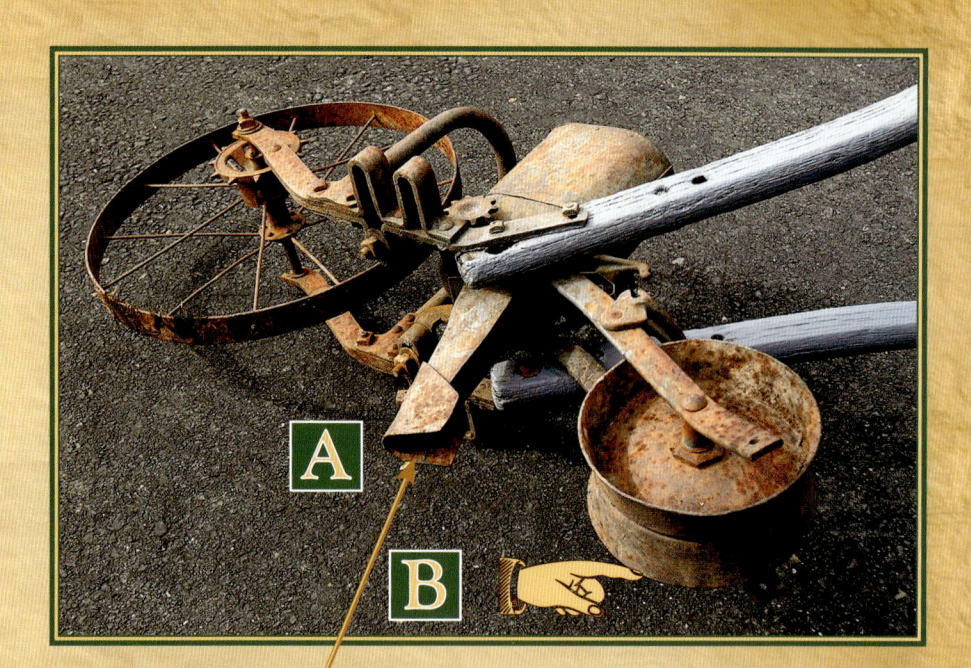

Avery Sure Drop Planter

B.F. Avery & Sons
1884-1916

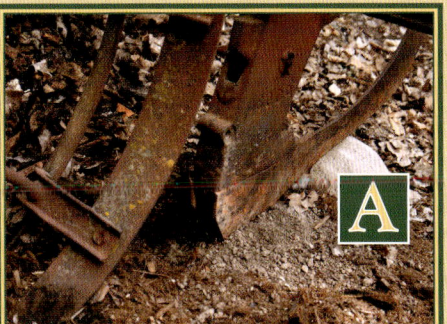

This implement is one of the first seed planters used to plant rows of plants instead of individual hills. Open-heel runners formed furrows into which dual seed canisters dropped seeds. Seed spacing in furrows was controlled by a wheel chain rotary plate under each canister that was preset for drop distance. The open press wheels behind the seed drop pressed the seed into the furrow.

This unit had two check-row markers that showed the distance between the already planted row and the next row to be planted. The runners, seed canisters and wheels could be adjusted for row widths. The operator used gear levers to raise the unit out of operation and to control furrow depth. Seed covers were spring loaded, and press wheels were equipped with scrape cleaners.

A. Open-heel runners
B. Seed canisters
C. Rotor plates system
D. Open press wheels
E. Check row marker
F. Gear control

Grain or Seed Drill
Brantingham Co.
1923▸

Disc opened furrows in front of the seed drops. A drill at each seed chute opening buried seed as it was pushed into the soil.

A. Wooden box for seed would have been on top of seed drops
B. Seed chutes and discs with scraper blades

Seed Chute

Scraper Blade

12

Keystone Corn Planter

Keystone Manufacturing Co.
1867

Sitting in the front of the planter, the driver controlled the horses and operated the width lever control while another man sat in the back and managed the seed drop operation. A 45-degree angle rotating grooved disc fed seeds from a large seed hopper into a rear grooved flat disc that controlled seed spacing by dropping seeds into a drill. The seed discs were powered using wheel axle gears. Furrows were created in front of the seed drop drill by an up-front single plow blade and tool to deepen the furrow. The angled discs in the rear covered the seeds with soil.

A. Front and rear seats
B. Seed hopper
C. Rotating grooved seed plates
D. Furrow adjustable plows
E. Rear discs and seed drop

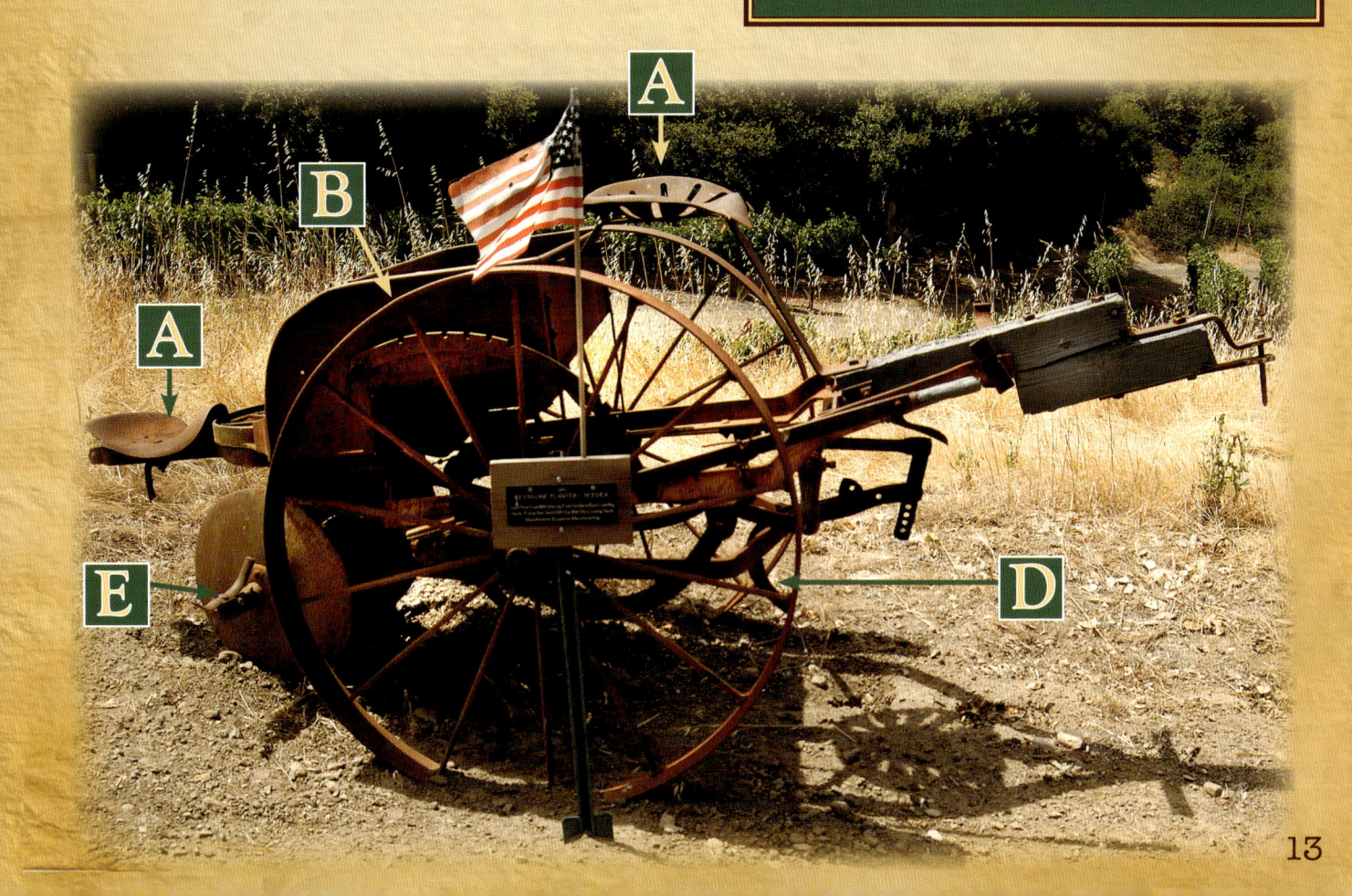

13

Check Row Marker
Briggs & Enoch Mfg.
◂1899▸

A check marker pulled across a field lengthwise and then crosswise, forming a square that showed the exact location for planting. The marker rotated on an overhead frame, allowing it to mark on either side for row or hill planting.

By removing seed canisters, this unit may have been modified from a combined seed planter and row marker.

A. Dual runners
B. Row width adjustments for wheels and runners
C. Marker controls
D. Check marker

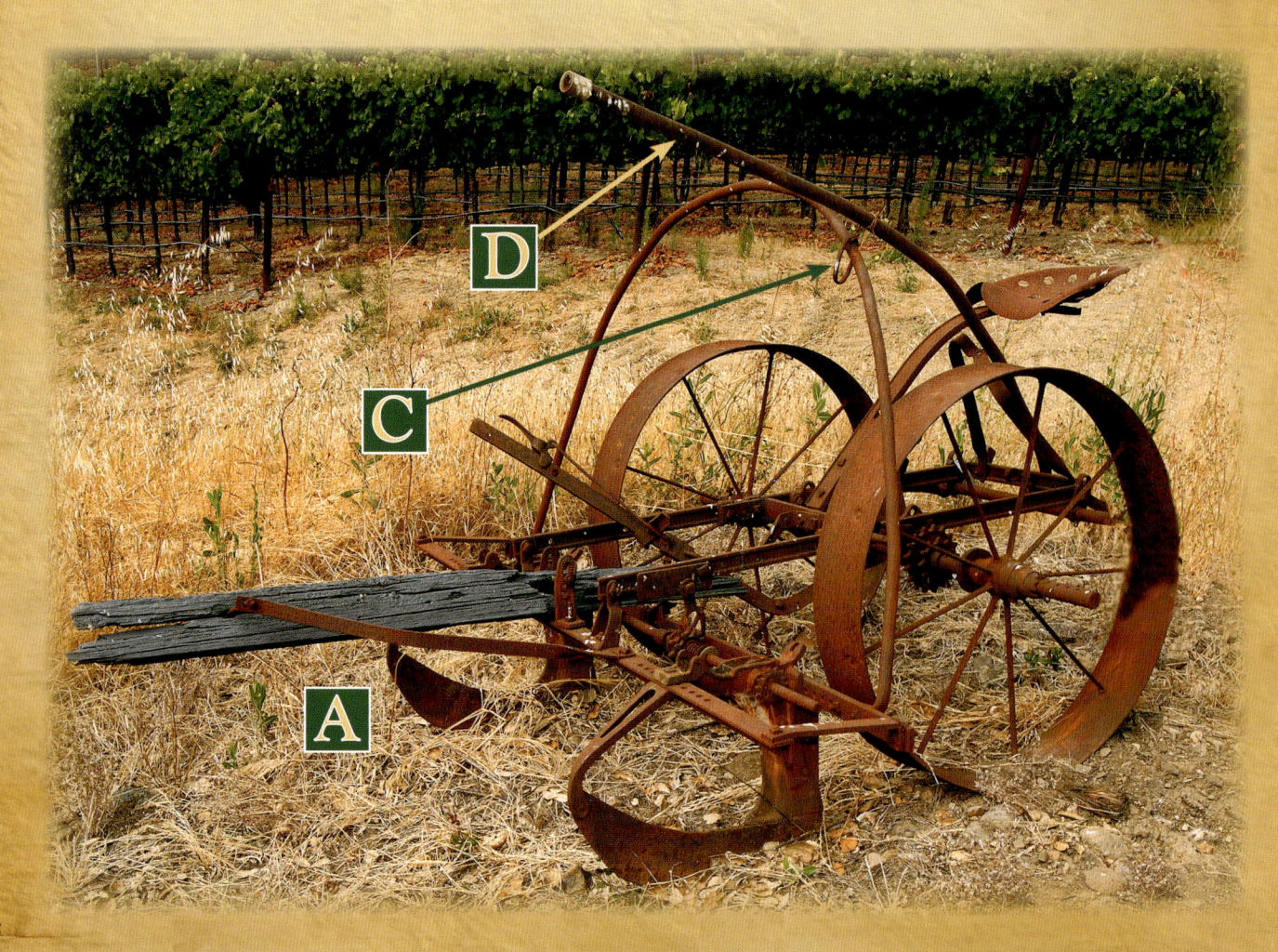

Cultivators

Dual Purpose Cultivator

Brown Man Co.
Date Unknown

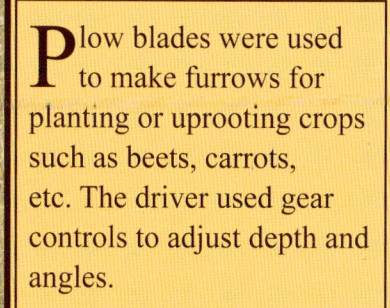

Plow blades were used to make furrows for planting or uprooting crops such as beets, carrots, etc. The driver used gear controls to adjust depth and angles.

A. Uprooting blades/depth control
B. Angle controls

15

No. 101 Planet Junior
S. L. Allen & Co
◁1905▷

Sometimes called a "horse shoe," this implement was used for light-duty weed tilling and furrowing. It has seven shovel-type blades, but they can be changed out to other types of cutting tools, such as sharp teeth or cutting blades. The 101 Planet Jr. was very popular for small farms and vineyards, and was suitable for truck farming. A lever expander adjusted cultivating width.

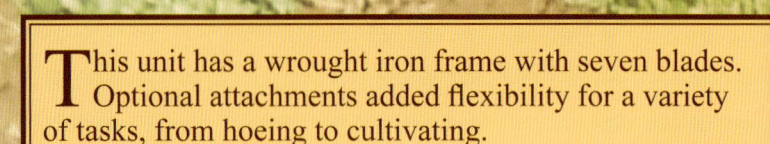

This unit has a wrought iron frame with seven blades. Optional attachments added flexibility for a variety of tasks, from hoeing to cultivating.

D. Lever allowed expansion to enlarge cultivation width.
E. Hookup options
F. Plow blades

A. The front-end connection shows how blade depth could be controlled by the position of the front wheel and height of the pulling connection.
B. Unit's width could be expanded depending on the row width desired. Notches show the lever control options.
C. Cutting blades

16

Garden Cultivator
Case-Keeler
1847 ▸

The Garden Cultivator was pushed by hand in small gardens. Optional attachments included the combination rake, plow, disc, and other implements.

A. Cultivation fork and furrow blade

Garden Cultivator
Peter Henderson & Co.
1847 ▸

A. Blade
B. Tools to substitute

17

Planter's Pride
Ohio Cultivator Co.
◄1905►

Shown without dual handles, this diamond-tooth harrow and cultivator's steel teeth pulverized and loosened soil without throwing dirt on plants.

A. Hookup angle adjustments
B. Lever to adjust cultivating width

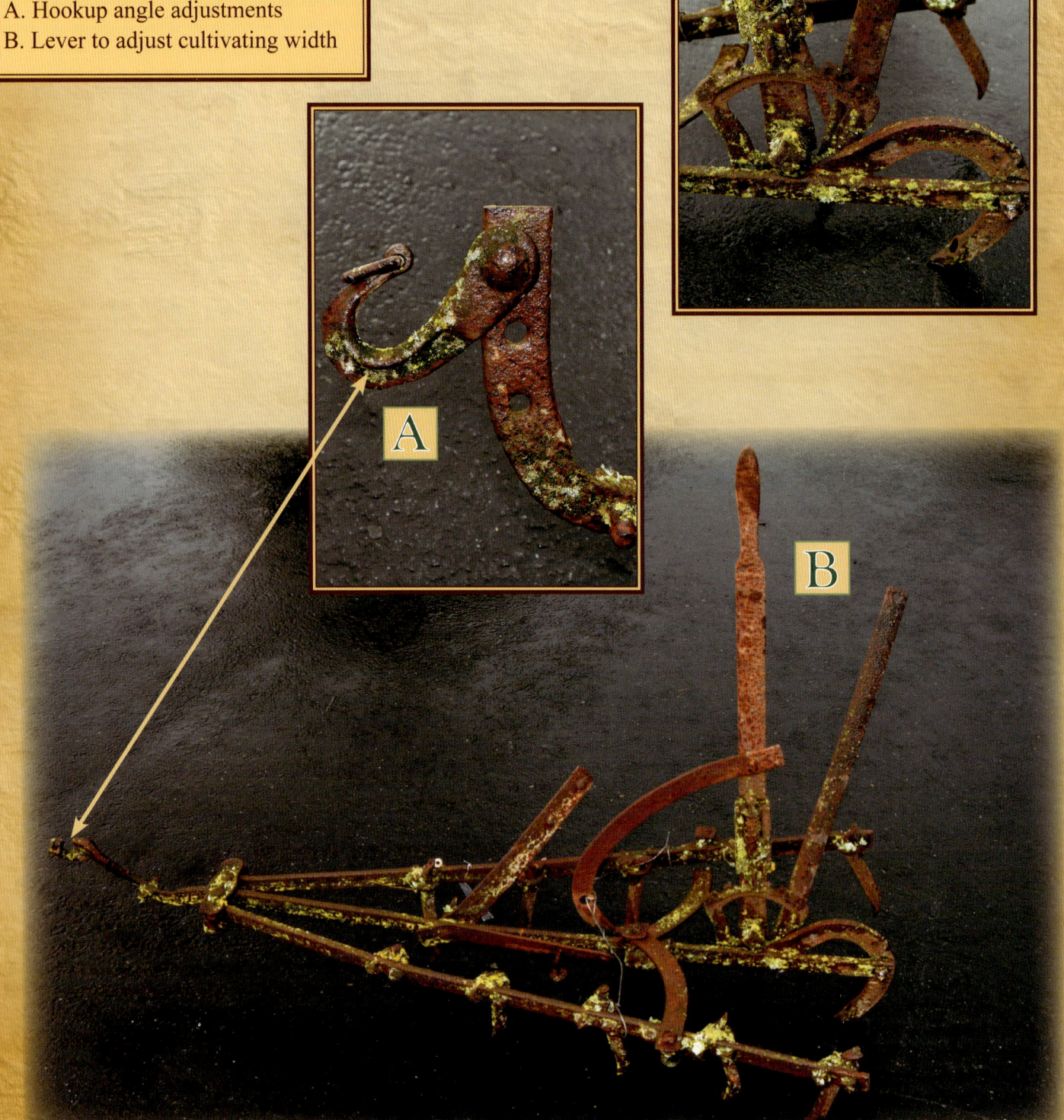

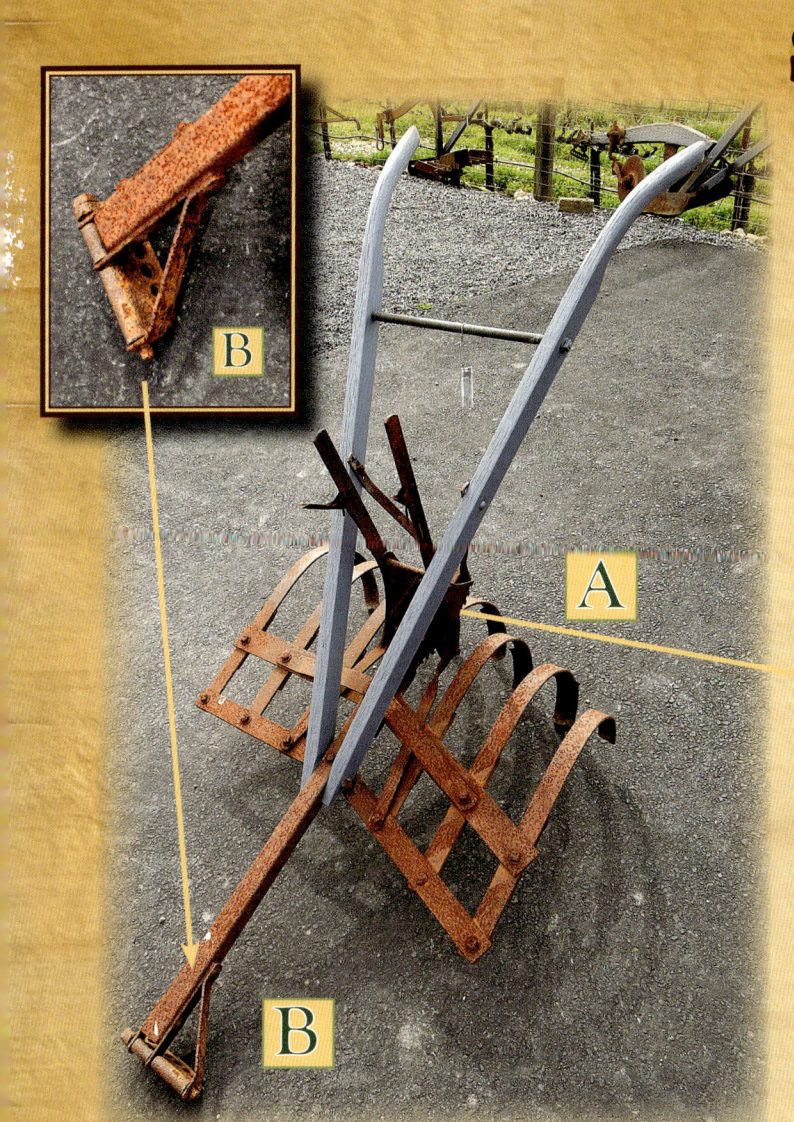

Surface Cultivator
Deere & Co.
◀1900▶

This horse-drawn implement was sometimes called a "Scratcher Spring". The teeth lightly scratched the soil, removing small weeds.

A. Dual height and width control mechanism
B. Hookup option

Double Fork Rake
Case-Keeler
1847▶

The Double Fork Rake was pushed by hand and was effective for shallow cultivation in small gardens.

Jack Rabbit Cultivator

B.F. Avery
◁1905▷

Four horses were used to draw this Jack Rabbit Cultivator when attachments called "pipe beam gang shanks" were attached. This Jack Rabbit Cultivator was used on the Elsbree Ranch for many years.

A. Each lever controlled half of the cultivator's depth.
B. Tool holders shown do not have cultivator implements attached.

Buckeye Cultivator
American Seeding Company
‹1915›

A variety of implements could be added to this gang-shank cultivator, which was pulled by four horses. This Buckeye Cultivator was used on the Elsbree Ranch in Windsor, California.

A. Adjustable tine mechanism
B. Depth adjustment mechanism
C. Single tree hookups

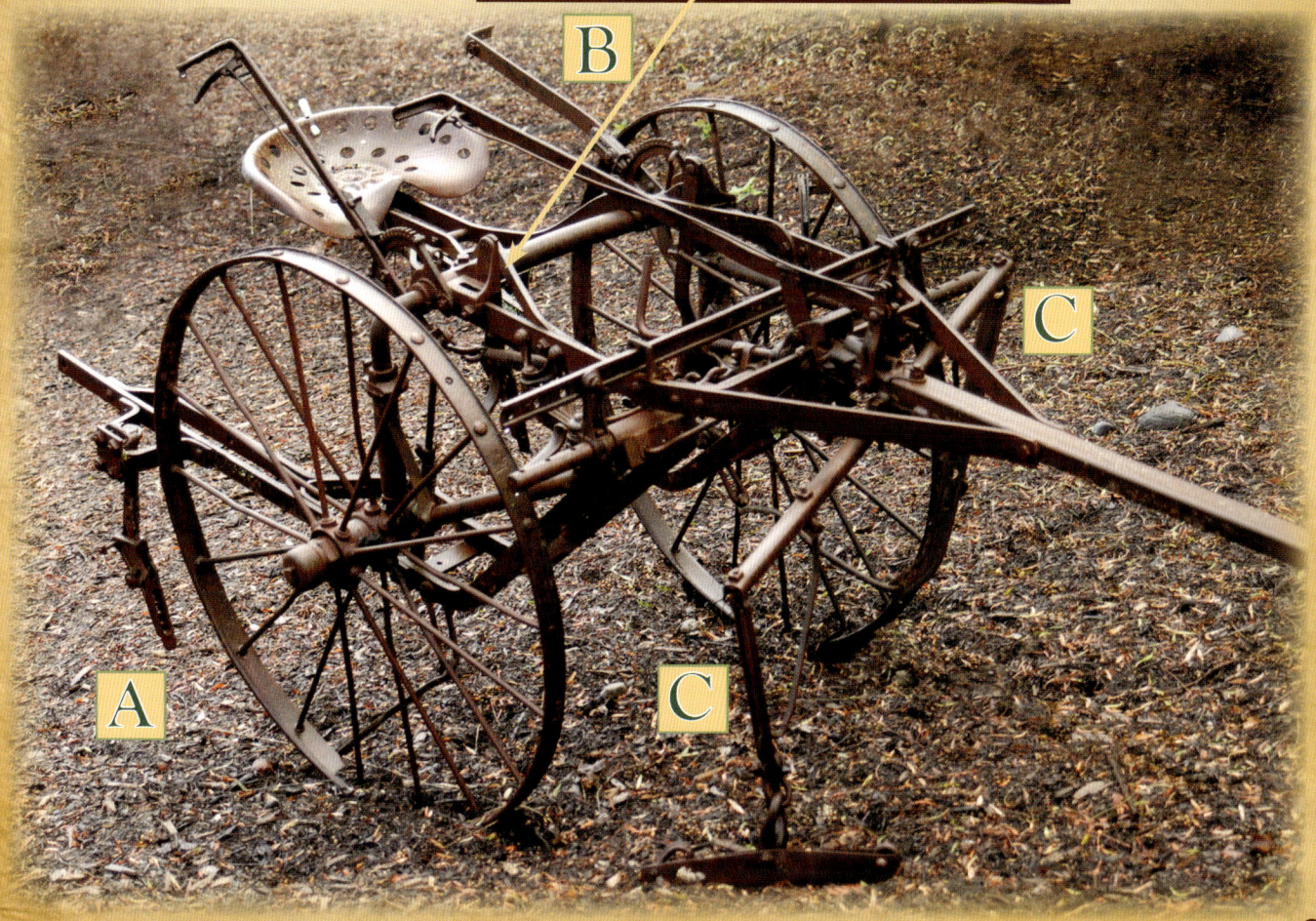

Small Cultivator
S.L. Allen & Co.
1890

This Small Cultivator has the same features as most Planet Juniors. It could expand plows to different row widths and various attachments could be used for working the soil.

A. Front-end connection for depth control
B. Lever for expanding width
C. Plow blade

Complex Cultivator
Case Plow Works
‹1920›

When this Complex Cultivator was first used on the Elsbree Ranch, it was horse-drawn. Later, it was converted for tractor power. A cultivator would normally use five implements to break up ground and make two furrows for planting.

Close up view of the lever controls

A. Blade and tines — shows spring load to ease equipment damage
B. Levers controlling depth and angles

23

Belanger Steel Scuffler
A. Belanger Ltd.
◁1867▷

This Belanger Steel Scuffler, which resembles the Planet Jr., is shown with seven shovels for softening and weeding the soil. Interchangeable implements could easily be attached for other uses. The width was expandable and the front-end connection options allowed for depth control.

A. Width control
B. Depth control
C. Plow shovels

Cultivator
A. Belanger Ltd.
1910

A lever expander was adjustable for row width. This unit shows that various attachments could be used.

A. Connection piece with adjustment options for depth control
B. Lever expander

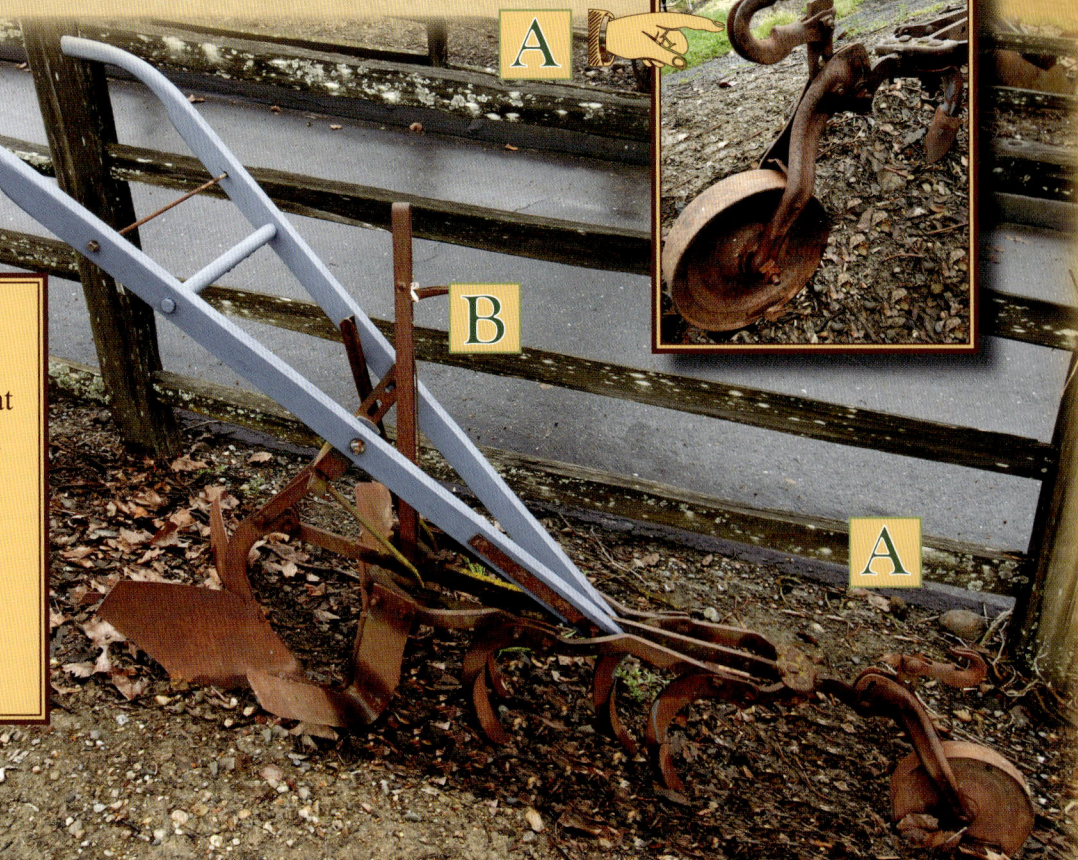

No. 101 Planet Junior Horse Hoe

Peter Henderson & Co.
1847

With this tool's changeable implements, the farmer could hoe and cultivate. The unit shown has a wrought iron frame with small front shovels to soften and weed the soil while the large back shovel created a furrow for planting. The frame could be expanded to a desired width, up to four feet.

A. Adjusting lever for expanding frame
B. Front-end connection options for depth control
C. Plow shovels

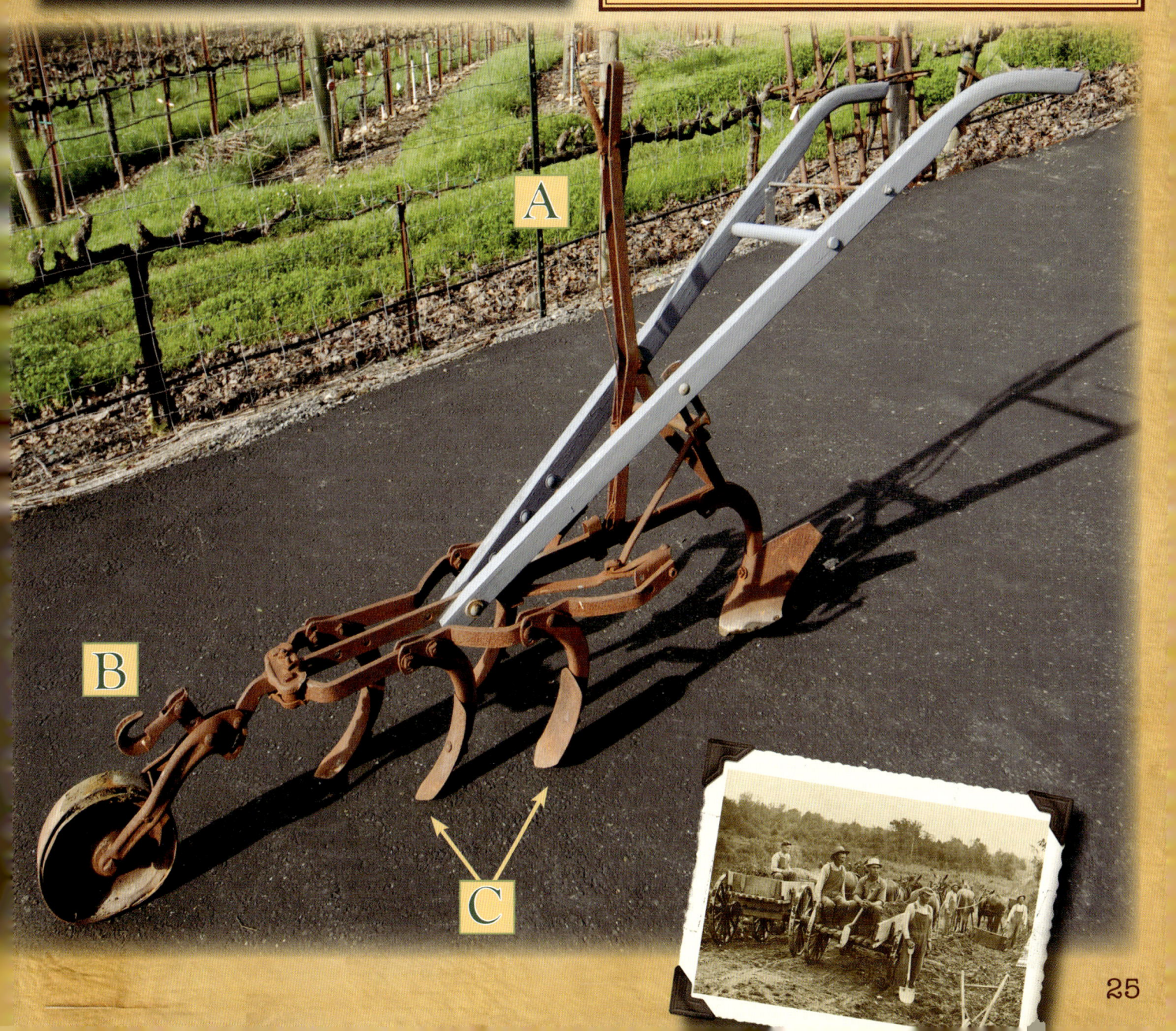

25

Harrows

Spring Tooth Harrow
Moline Plow Co.
1887-1915

This one-horse vineyard harrow was used for leveling and smoothing soil, especially stony ground. The spring teeth could be adjusted for the angle required for the roughness of the ground's surface. A slide plate was attached to the front corners to keep the frame floating on the soil's surface.

A. Gear and handle to position spring teeth
B. Ground surface plate
C. Hookup options to control depth
D. Spring teeth shown in up position

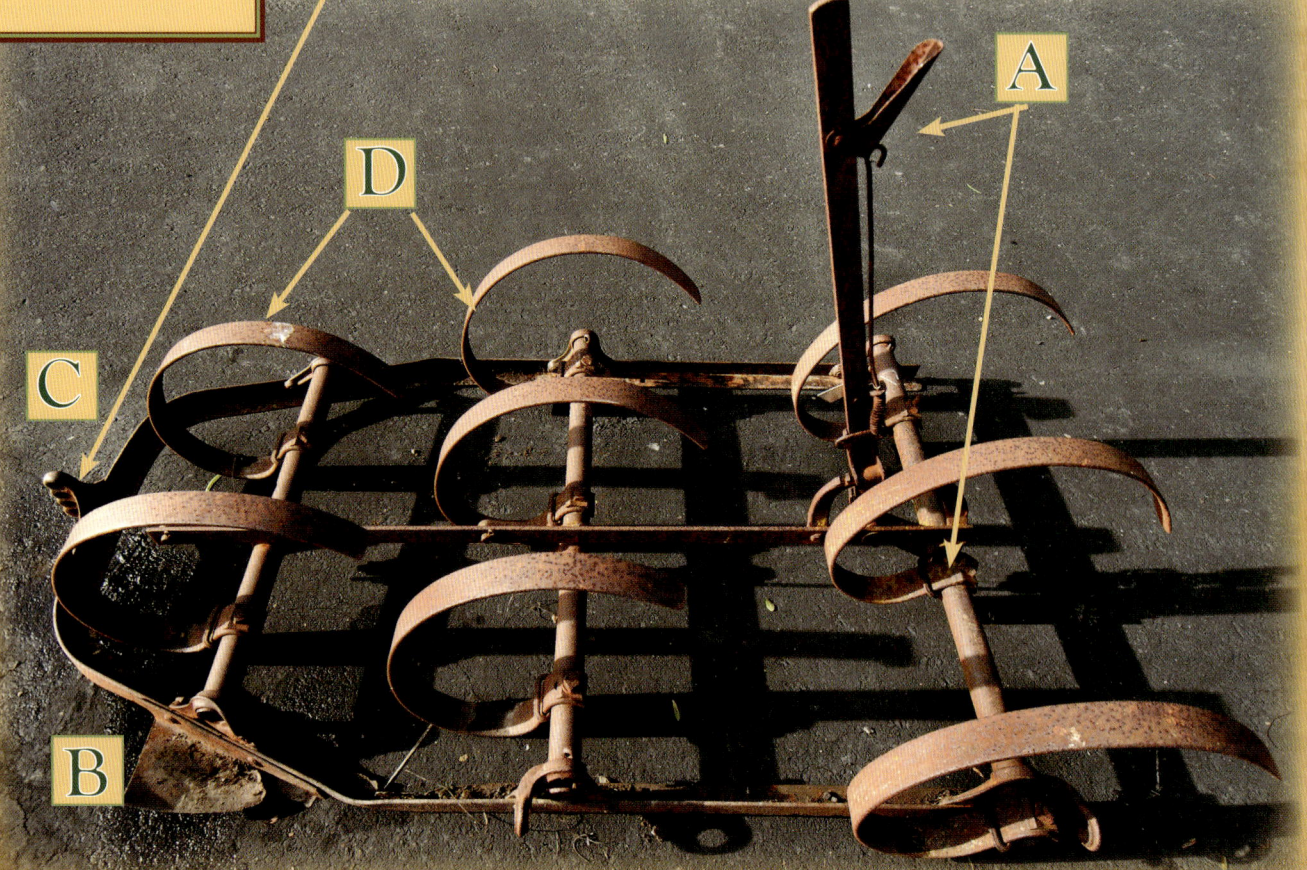

Revolving Harrow
Fresno Agricultural Works
1870-1900

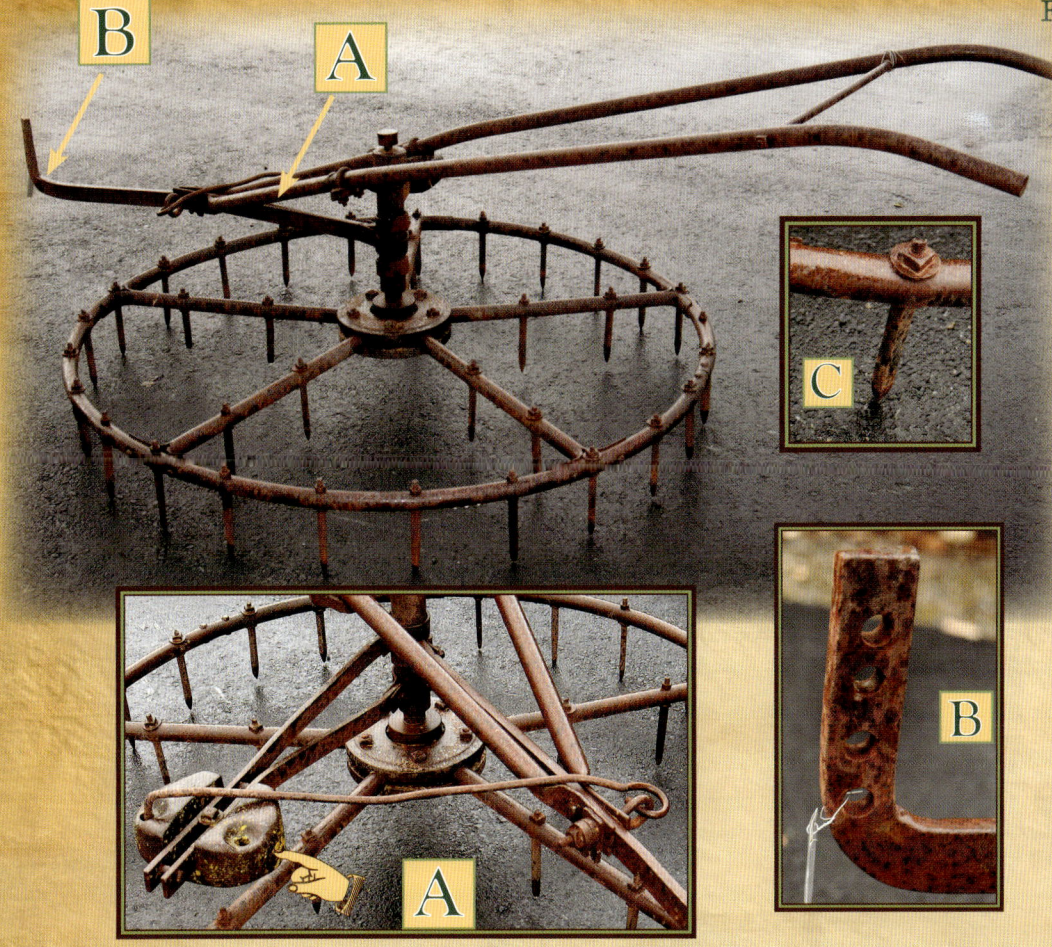

This tool was used primarily in vineyards to cultivate close to grapevine trunks without damaging them. It was also used for leveling. The rotary spikes revolved as a horse team pulled the device and the operator guided it using the handle and the rudder to obtain a straight line.

A. Rudder lock
B. Front end connections options for depth control
C. Spike tooth

A-Frame Harrow
Avery & Sons
1878

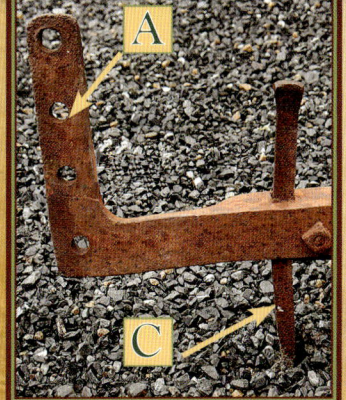

This harrow with 24 spike teeth was used primarily in vineyards. The width could be adjusted to actual vine row spacing. It allowed for aggressive cultivation because the A-Frame design prevented hang-up on trunks and other obstructions.

A. Hookup and height adjustment options
B. Width control mechanism
C. Spike tooth

Ten Blade Disc
Emerson-Brantingham Co.
◁1907▷

This implement included five disc blades for each side with dual curvature controls. By adjusting the angle of the disc curvature, the amount of soil pulverization could be controlled. The Ten Blade Disc replaced moldboard plows.

A. Cutting disc blades
B. Dual adjusting levers for disc's curvature
C. Depth adjustment lever
D. Oil fittings
E. Weight racks

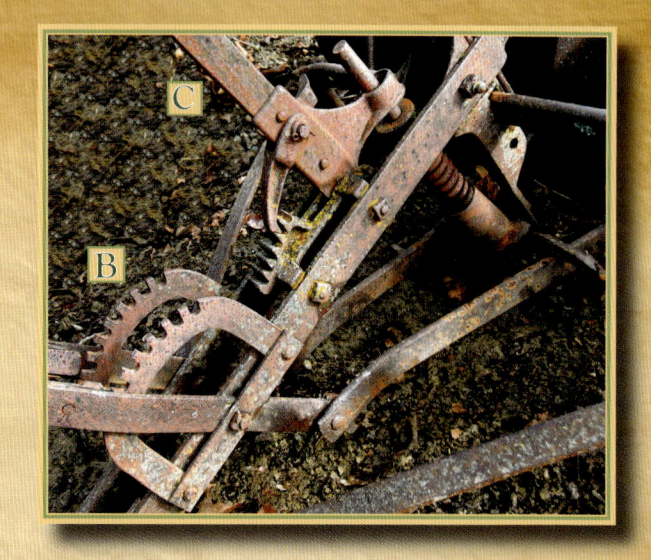

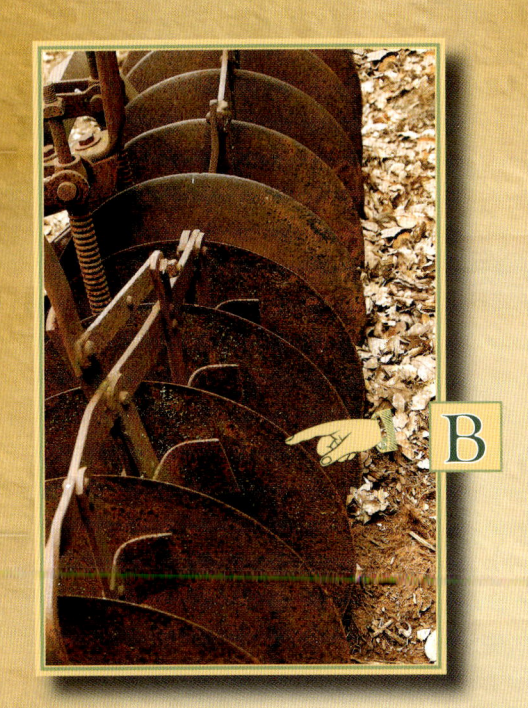

Pennsylvania Disc Harrow

Deere & Co.
1905

Model B Disc Harrow

This ten-foot Disc Harrow has five blades on each side. The driver controlled an oscillating scraper system mounted on each disc to keep them clean, and using lever controls, he could change the pitch independently for each set of five discs. The primary use of the Disc Harrow was for tilling weeds and softening surface soil

A. Pitch lever control   B. Disc scraper system

Harrow Sulky
By David Bradley Mfg. Co.
◀1900▶

Different types of harrows could be attached to the draw bars of the sulky, which allowed the operator to ride rather than walk behind. The protruding bar at the rear of the sulky suggests that an implement could also be dragged behind.

Iron Age Harrow
Peter Henderson & Co.
‹1847›

This single wheel harrow was pushed by hand and was ideal for use in small gardens.

A. Harrow element

Harrow Cultivator
Peter Henderson & Co.
◁1900▷

This cultivator has a solid wrought iron frame with diamond teeth, and it was used primarily as a garden and field weeder. The frame can be expanded or contracted for row width.

A. Hookup adjustment with single tree
B. Adjusting width mechanism
C. Diamond tooth

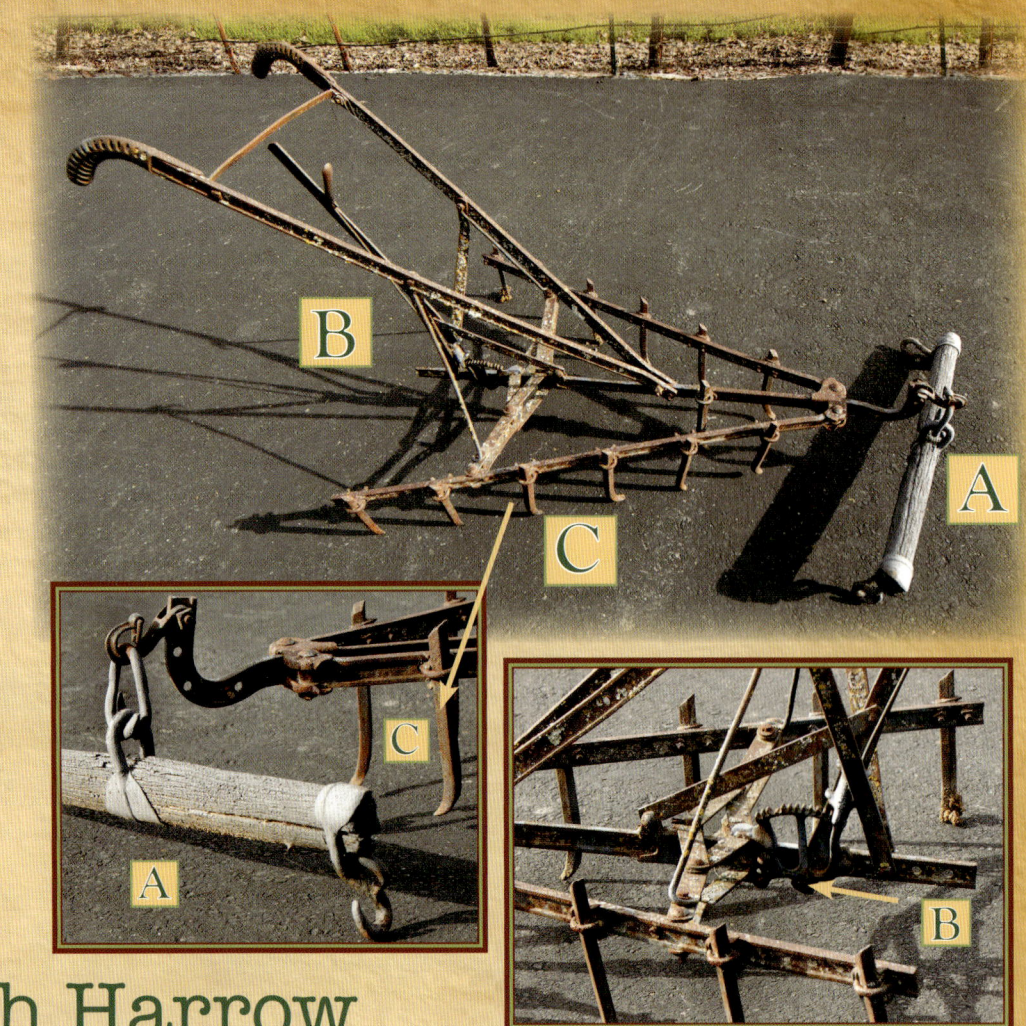

Spike Tooth Harrow
Bucher & Gibbs Plow Co.
◁1900▷

This harrow was used for aggressive leveling and smoothing of soil. For more aggressive work, the spikes could be set in a vertical position. For less aggressive work, spikes could be set almost flat.

A. Lever control for spike pitch
B. Front end hookup
C. Spike tooth

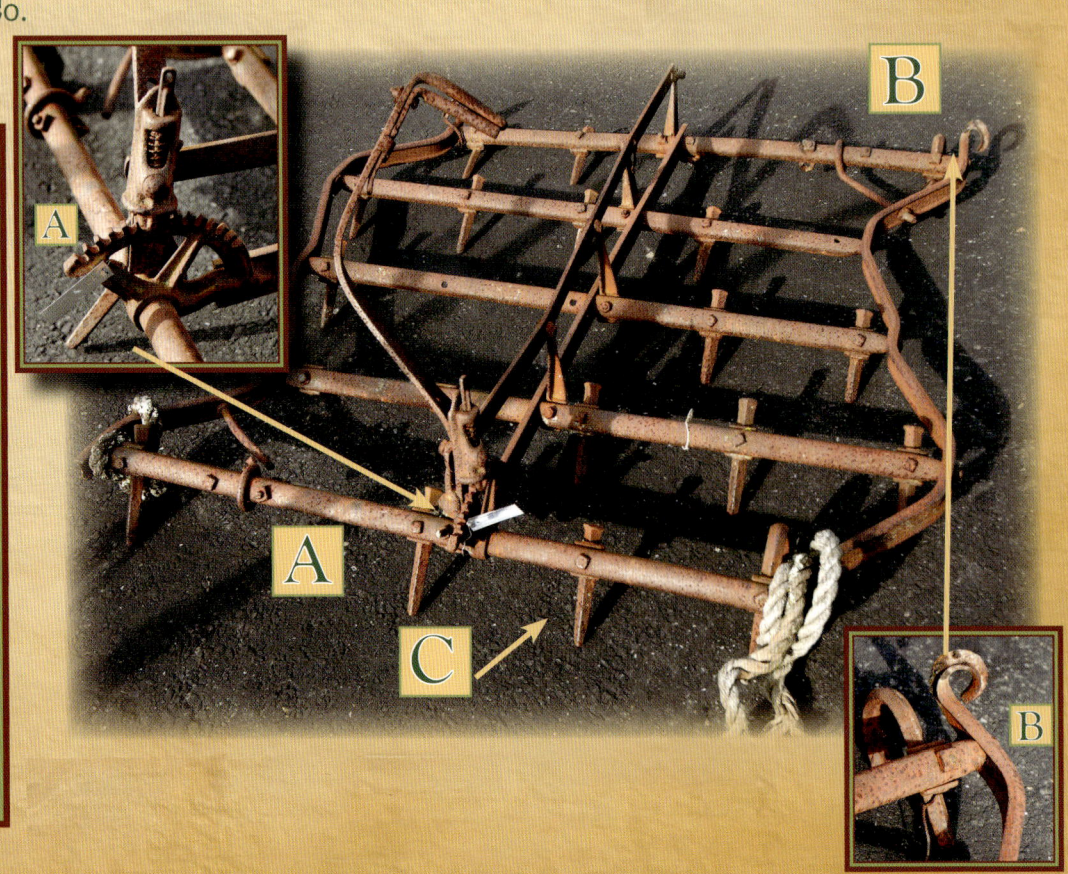

Plows

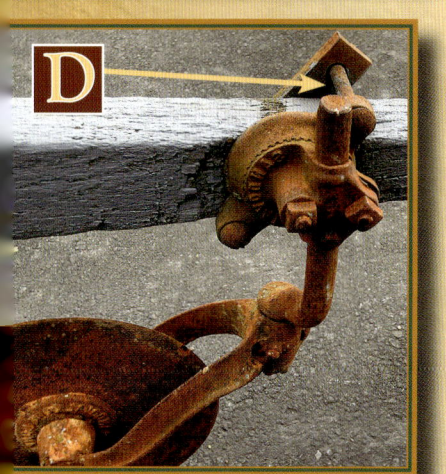

Prairie Breaker Plow
Brantingham Co.
◁1880▷

The Prairie Breaker was able to cut through the tough roots of prairie sod, reaching depths of 16 inches. Its coulter—a cutting blade or wheel— made a vertical cut in the soil ahead of the plowshare and helped keep the moldboard clean. The moldboard lifted the sod and rolled it over in one long strip, covering vegetation.

Most of the horse-drawn plows have all or some of the following features:
A. Wood or metal handles
B. Wood or metal beam
C. Double tree attachment
D. Coulter
E. Plow Share
F. Mold Board
G. Plow angle adjustment

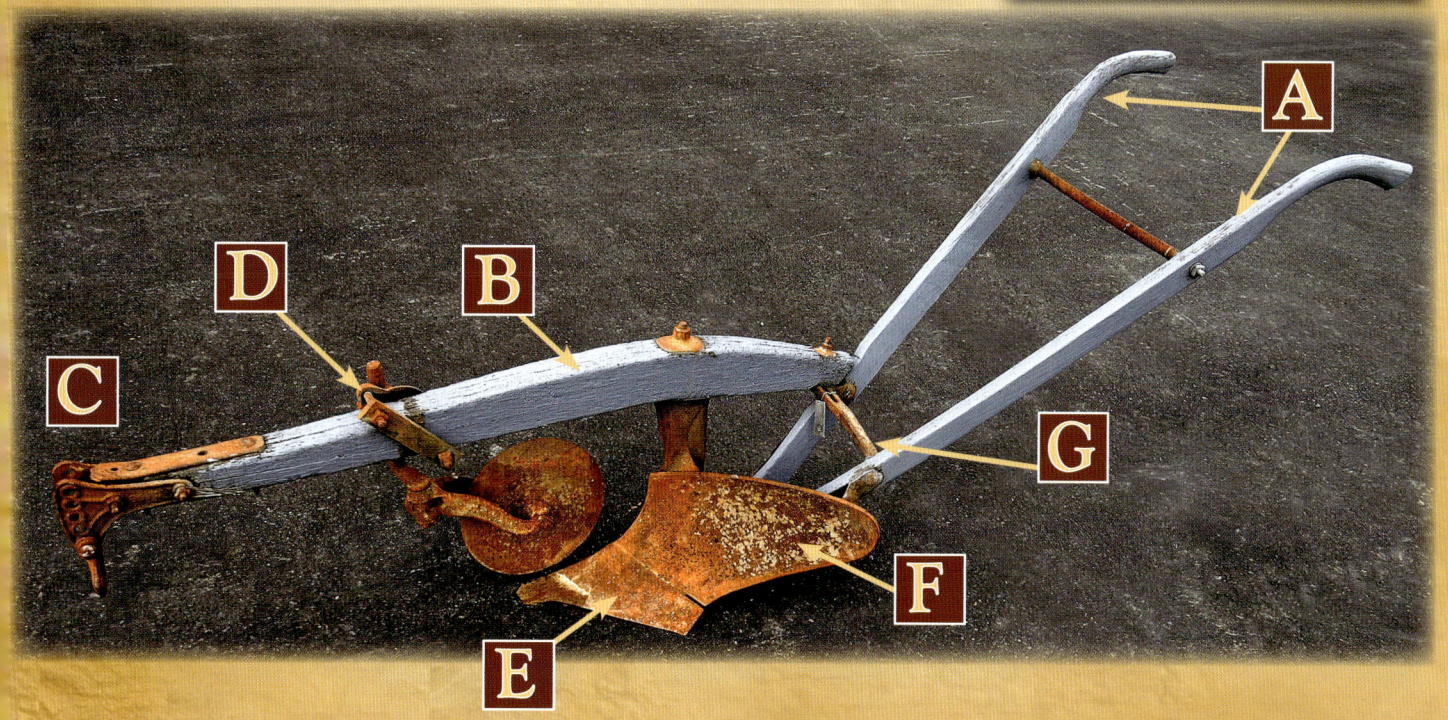

33

Wooden Double Shovel Plow

Avery & Sons
‹1897›

The Wooden Double Shovel Plow is one of the oldest known designs, an improved version of the single shovel plow. The beam connection to a single or double tree could be adjusted for angle.

A. Angle adjustments
B. Plow shovels

Walking Furrow Plow

Oliver Chilled Plow Works
1892

The Walking Furrow Plow was normally a dual-horse plow with adjustable angles. It was very common and very inexpensive.

A. A wide range of hookups allowed for depth and angle options.
B. Angle adjustment

Canton Clipper Plow
Parlin & Orendorff
1842

Wide and deep cutting, the Canton Clipper Plow required a team of horses.

A. Depth control options

Reversible Plow
B.F. Avery
◁1900▷

The primary advantage of this implement was that when plowing a hillside the moldboard could be reversed to throw the furrow slice downhill regardless of the direction being plowed, which helped prevent soil erosion. Reversible plows were later replaced by two-way sulky plows.

A. Hookup options for pulling right or left angles and for depth of cut
B. Locking clevis for shifting moldboard to left or right depending on which direction the horse team was heading

35

Red Jacket Walking Plow
Oliver Co.
1880-1900

This was a chilled-iron plow, made with iron that had been cooled quickly with a stream of water to increase the metal's hardness. The process was perfected by a Scots immigrant, James A. Oliver. Farmers also called this implement a "stirring plow."

A. Angle adjustments
B. Hookup options
C. Moldboard
D. Plowshare

Furrow Plow
Lacrosse Plow Co.
◄1900►

This plow is offset from the frame and the operator. The angle of the plow allowed the operator to remove dirt buildup next to grapevines and orchard tree trunks.

A. Blade with offset angle
B. Wheel adjustment mechanism for depth and angle

36

Stirring Plow
John Deere
‹1900›

This all-metal plow was an essential piece of farm equipment. If a farmer had only a Stirring Plow, he could handle most necessary plowing tasks. Angle and depth adjustments could be set in the front and rear of the plow.

A. Chain angle adjustment
B. Handle pivoting adjustment

Double Mold Plow
Racine-Sattler Co.
‹1851›

This plow was used for opening large furrows for planting, hilling, ridging, trenching, and other tasks.

A. Rear view
B. Front hookup options

37

Ditch Irrigation Plow
Martin 1892➤

A. Plow share
B. Moldboard width adjustment
C. Hookup & wheel adjustment for depth

This heavy metal implement was used to create or clean out irrigation or drainage ditches on farmland. The plow share could be adjusted for angle, and the moldboards could be adjusted for width.

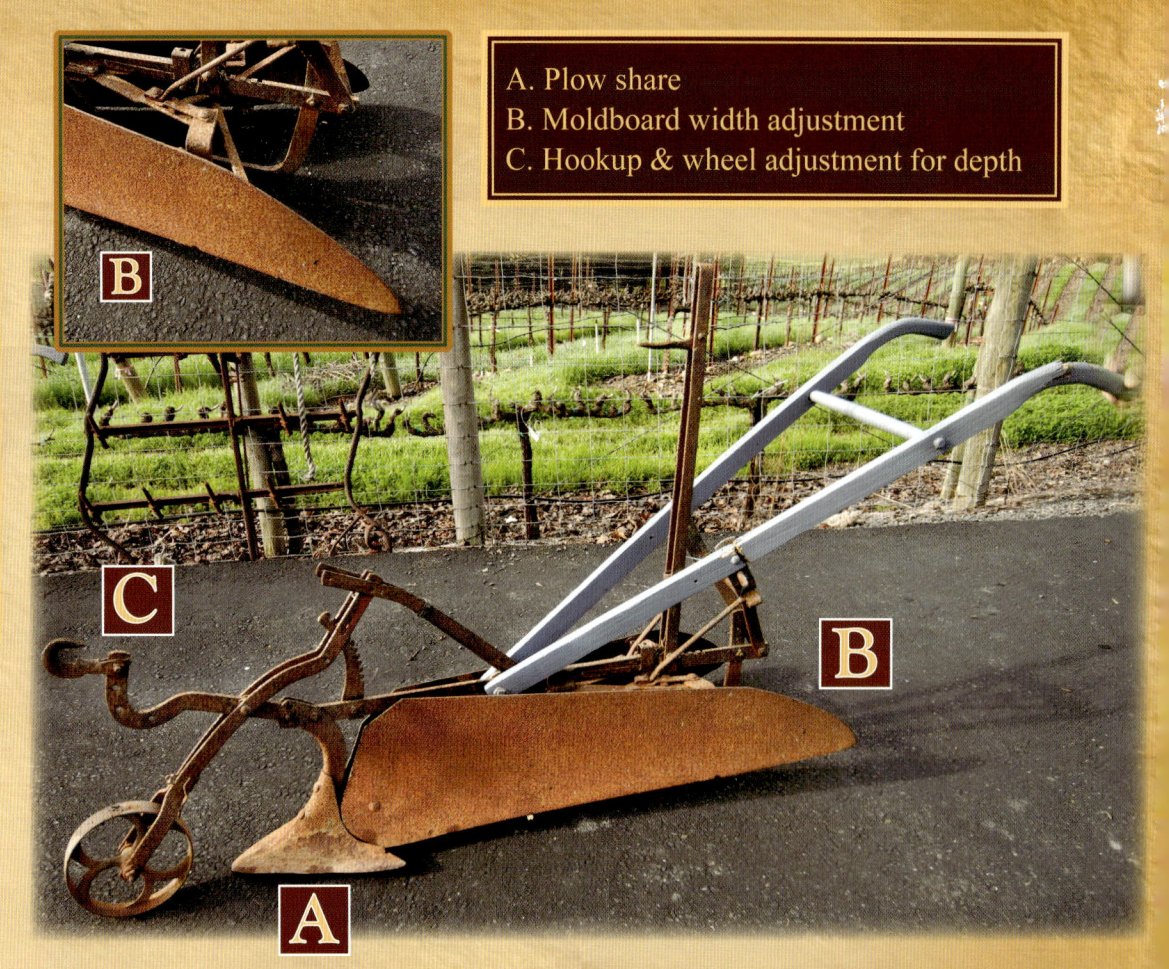

Ditch Cleaner Plow
Mogul 1892➤

Shallower plows were used for cleaning existing ditches; deeper plows, for digging new ditches.

A. Plow share
B. Moldboard width adjustment
C. Hookup and wheel adjustment for depth control

Potato Digger Walking Plow
Alexander Speer & Sons
1870-1910

This plow was pulled by a two-horse team and featured a dual-wheel front to maintain straight row digging. A wooden spindle divided and separated the potato vines. The large, flat-point shovel dug deep enough to uproot potatoes and bring them over the top of the vibrator tray, which separated dirt from the potatoes and left them exposed on the ground.

A. Revolving wooden spindle with its depth control
B. Front-end connection adjustments to help control depth and angle of digging
C. Vibrator or bumping tray

Potato Digger Plow
Martin 1847

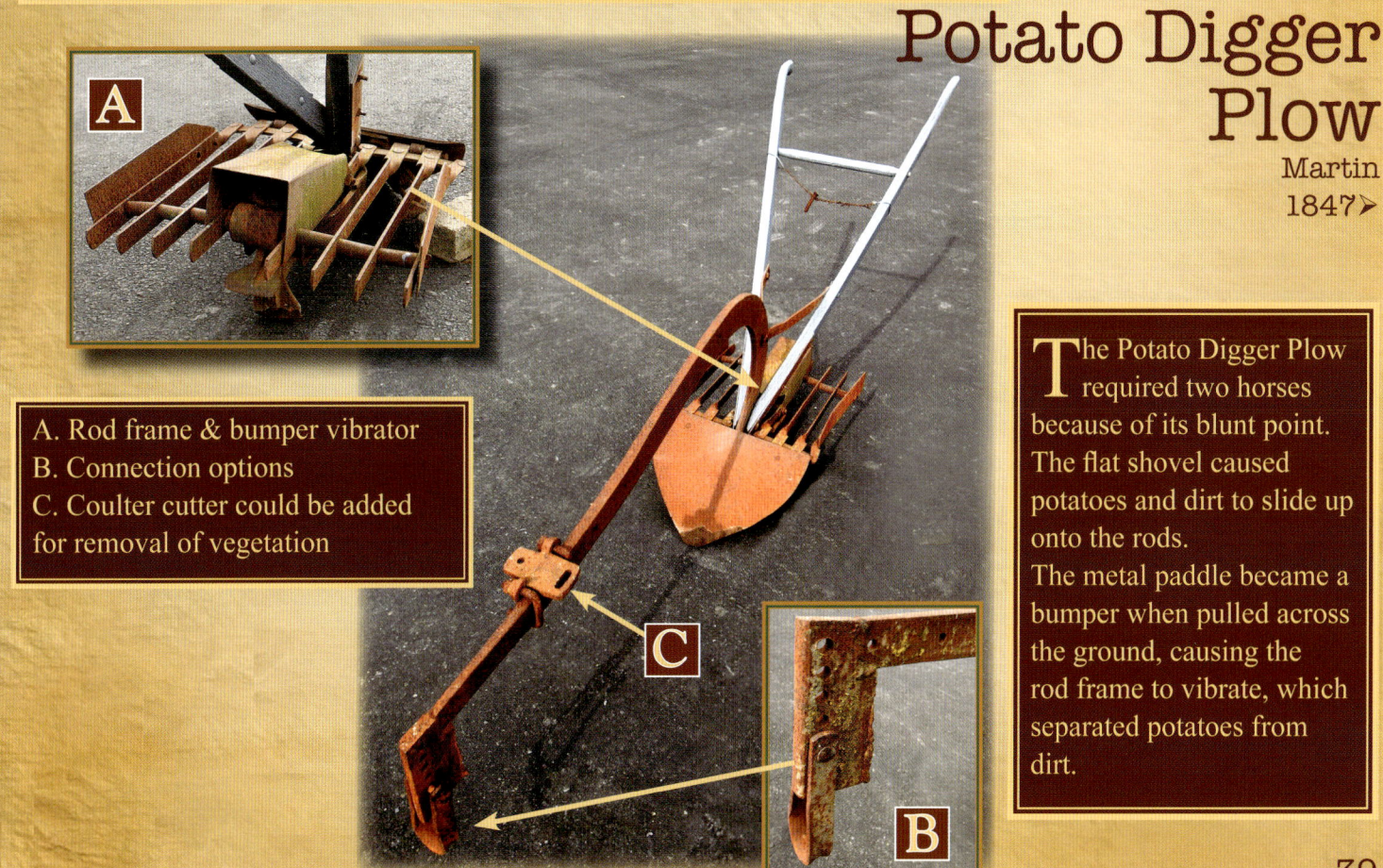

A. Rod frame & bumper vibrator
B. Connection options
C. Coulter cutter could be added for removal of vegetation

The Potato Digger Plow required two horses because of its blunt point. The flat shovel caused potatoes and dirt to slide up onto the rods.
The metal paddle became a bumper when pulled across the ground, causing the rod frame to vibrate, which separated potatoes from dirt.

39

Large Walking Plow
The Bryan Plow Co.
◄1900►

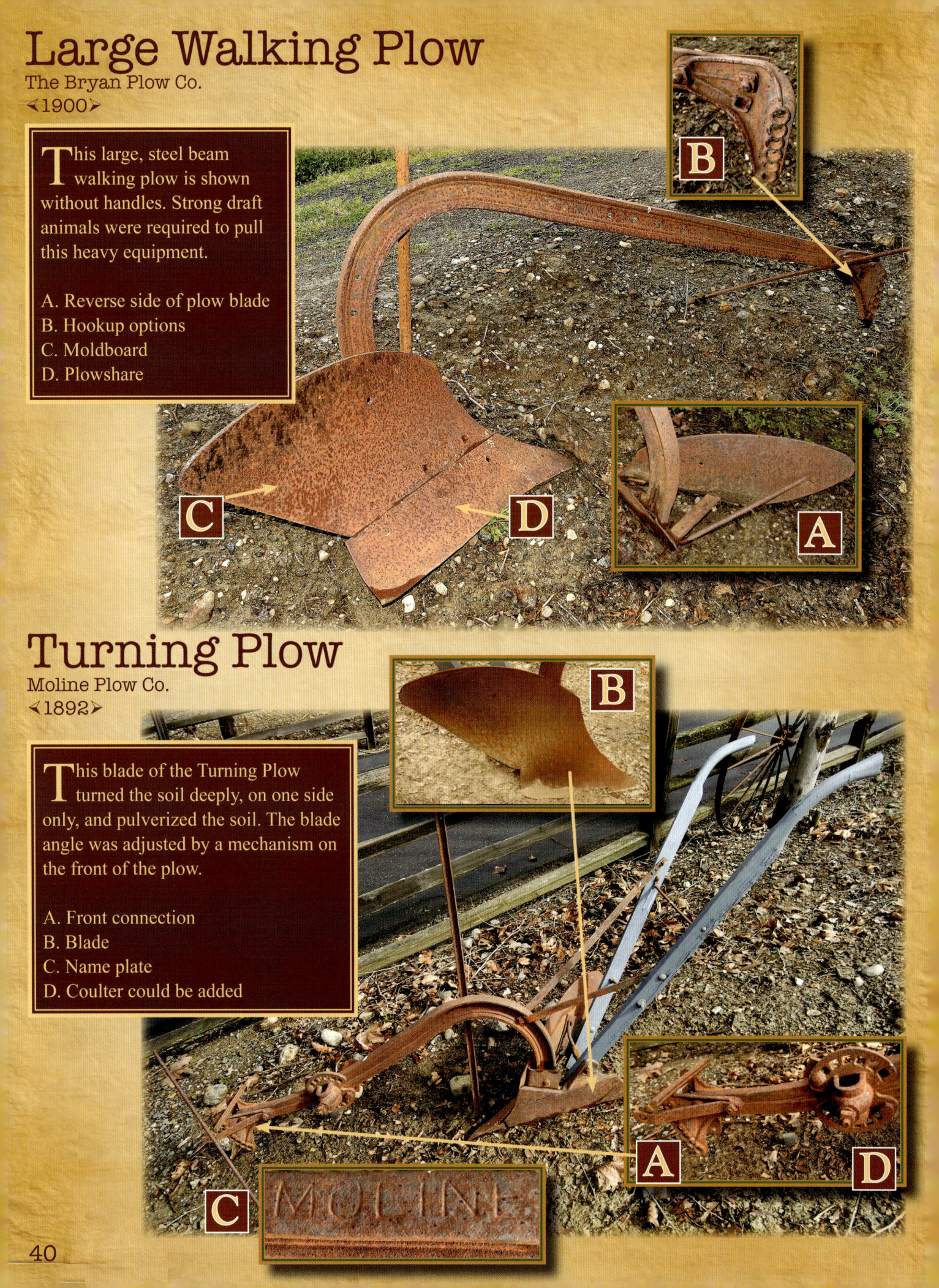

This large, steel beam walking plow is shown without handles. Strong draft animals were required to pull this heavy equipment.

A. Reverse side of plow blade
B. Hookup options
C. Moldboard
D. Plowshare

Turning Plow
Moline Plow Co.
◄1892►

This blade of the Turning Plow turned the soil deeply, on one side only, and pulverized the soil. The blade angle was adjusted by a mechanism on the front of the plow.

A. Front connection
B. Blade
C. Name plate
D. Coulter could be added

Improved Winged Shovel & Potato Digger Plow

Hoover Manufacturing Co.
1908

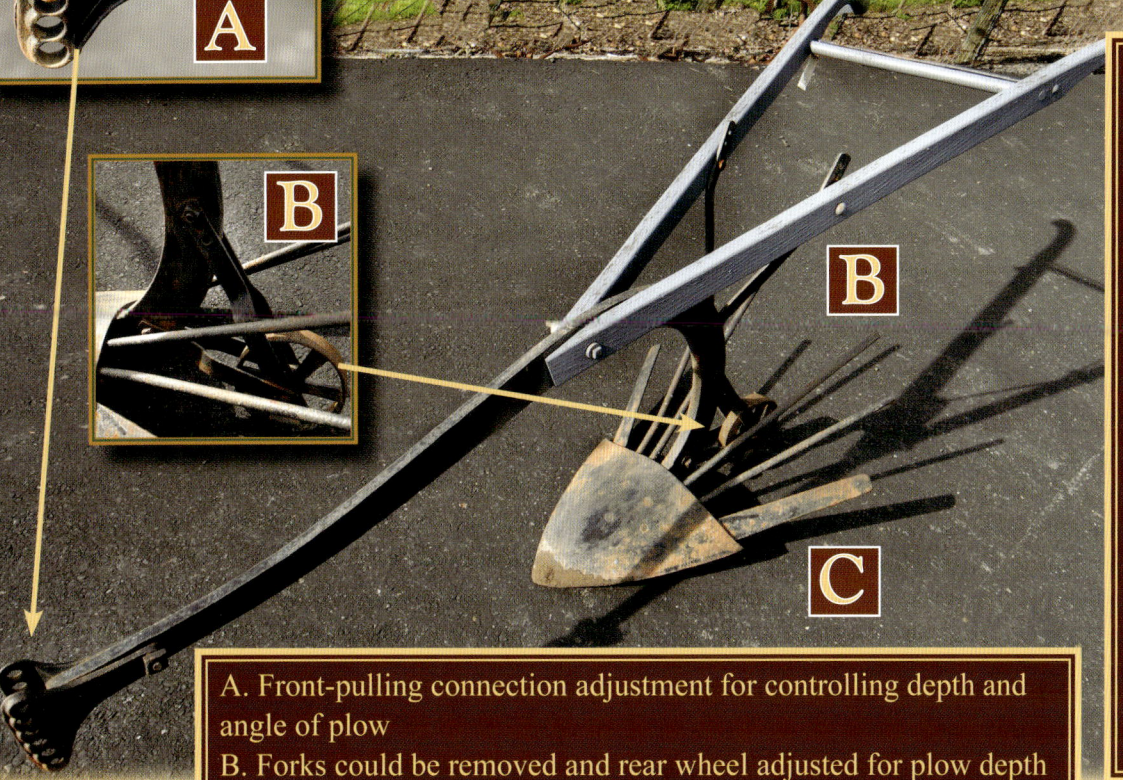

As a potato digger, the plow lifted potatoes from the ground and passed them over the forks to remove dirt.
As a winged shovel plow, the forks were removed and wings lowered for furrowing out or cultivating. The wings could be lowered to throw dirt against existing plants.

A. Front-pulling connection adjustment for controlling depth and angle of plow
B. Forks could be removed and rear wheel adjusted for plow depth
C. Wings

Vulcan Chilled Plow

Vulcan Plow Co./Southbend Chilled Plow Works
1897

The Vulcan Chilled Plow was used primarily for turning hard-rooted soil. Here, the plow is shown with a solid chilled moldboard.

A. Solid moldboard
B. Pulling connection adjustment for depth control
C. Adjustment for angle control

Oliver Sulky Plow
Oliver Chilled Plow Works
◂1892▸

There were left- and right-handed sulky plows, and this unit is a left-handed sulky plow, which always formed the upper side of furrows on the left side of the plow. The plow was equipped with control gears that lifted the plow out of operation for furrow depth and angle control. A tilt steering wheel allowed the farmer to make sharp or angled furrow turns.

A sulky plow was a metal frame with three wheels, a plow, and a driver's seat with controls. This was the first plow advancement over the moldboard walking plow. Soon after, a two-way sulky plow was developed with both left- and right-angled plows mounted in the framework. One plow could be lifted out of operation depending on the direction of plowing, which meant there would be no dead furrows.

A. Front-end connection
B. Joiner
C. Turning wheel
D. Coulter wheel
E. Scrape cleaner bar

Cambridge Reversible Plow

Lovejoy & Son
1906

The Reversible Plow was also known as a "hillside plow" because on hillsides, a farmer could reverse the plow to the left or the right, moving back and forth, and leave a furrow in the same direction, thus eliminating a dead furrow that might cause erosion on the downward side. Reversible plows were replaced by two-way sulky plows.

A. Front connection
B. Angle adjustment
C. Reverse mechanism
D. Position for coulter addition if required
E. Blade

43

Tractor Gang Plow
Massy-Harris Co.
1928 ➤

A. Crank adjustment for depth
B. Bull wheel
C. Till wheel
D. Plows

Two and three gang plows were added to frames as early tractor pulling power improved, and Case Plow Works was a developer of gang plows. Case was purchased by Massy-Harris and this gang plow is believed to be their model #28.

The plow frame could be lowered from the front wheels by a crank to control plow depth. The till wheel was also adjustable for depth. Tractors soon replaced horse teams for pulling heavy gang plows.

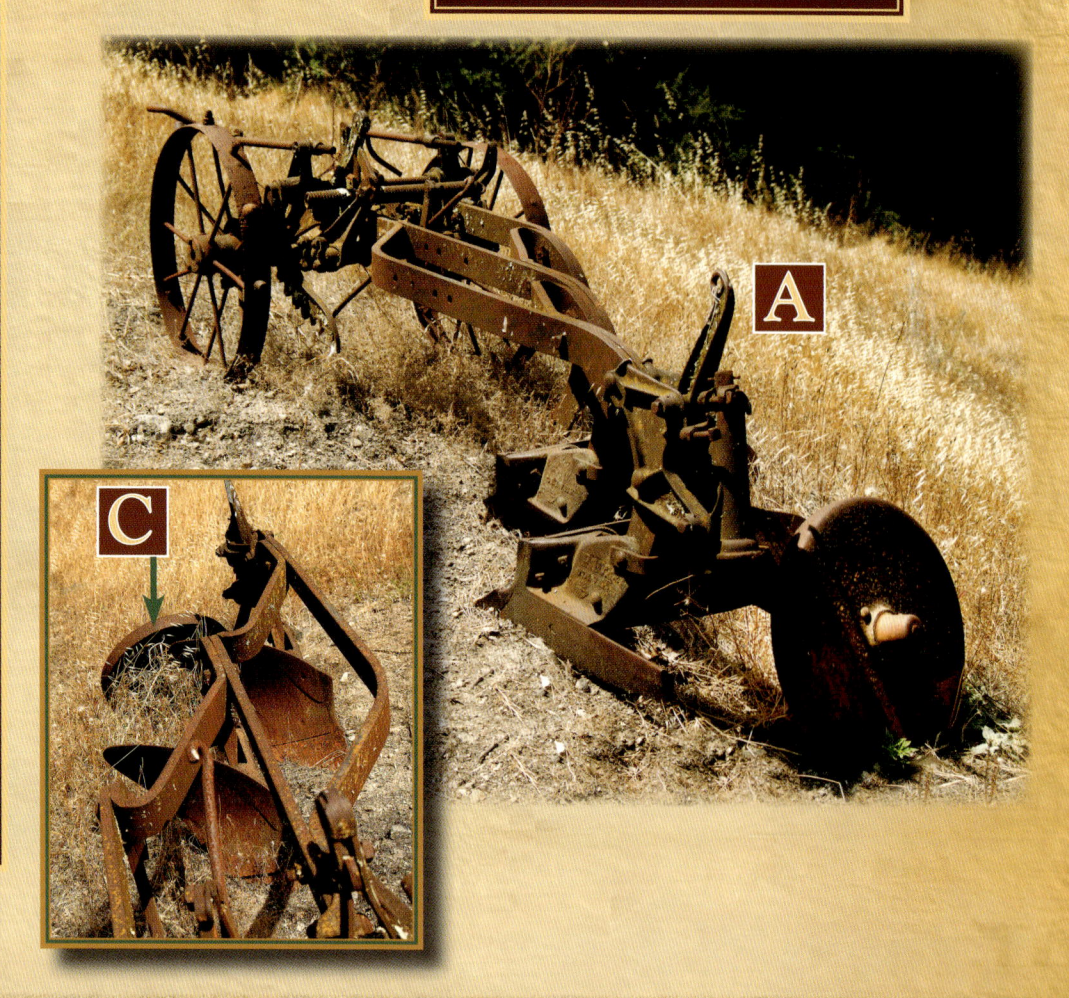

HAYING EQUIPMENT

California Grapple Fork
Whitman & Barnes 1900

The Grapple Fork replaced the hay sling for lifting hay into the barn for storage.

A. Connection piece for rope or hoist
B. Release lever to drop load
C. Grapple fork in closed position
D. Grapple fork in open position.
E. Spike to help secure the hay

Closed Position

Open Position

45

Hay Rake/Sling
F.E. Myers & Brothers
◂1890▸

The Hay Rake/Sling could be used as a large hay rake utilizing a wooden handle, but it was used primarily as a hay sling with a rope and pulley attached to an iron ring. Loose hay was then loaded into or unloaded from barns or wagons.

A. Pulley connection piece and hay release
B. Wooden handle

Hay Blower
Avery Co.
‹1920›

Hay was loaded onto a conveyor by hand. The conveyor fed the hay into flywheel blades, then the blower pushed the chopped hay into a silo or attic barn storage area.

A. Sharpening wheel for blades was built into the machine
B. Fly wheel cutter

47

Sandwich Hay Mower

Sandwich Manufacturing Co. 1892 ➢

A horse team pulled the Sandwich Hay Mower, which had a five-foot blade powered by a unique wheel chain-and-drive linkage.

A. Five-foot cutter bar
B. Heavy bull wheels drive gear box
C. Lever for blade control
D. Outer shoe
E. Inner shoe
F. Grass board

48

Ideal Mower
McCormick Deering
1800

A horse team pulled the Ideal Mower and its wheels powered the five-foot cutting blades. While most blade bars were four- to five-feet in length, eight-foot lengths were available but required larger horse teams.

A. Bull wheels
B. Five-foot cutter bar
C. Blade control levers
D. Outer shoe
E. Inner shoe
F. Grass board

49

Hay Tedder
Thomas Man Co.
◁1887▷

After hay was cut and lay on the ground, the Hay Tedder, which had moving forks, would be used periodically to turn the hay for uniform drying prior to raking in rows.

A. Spring forks
B. Engaging controls

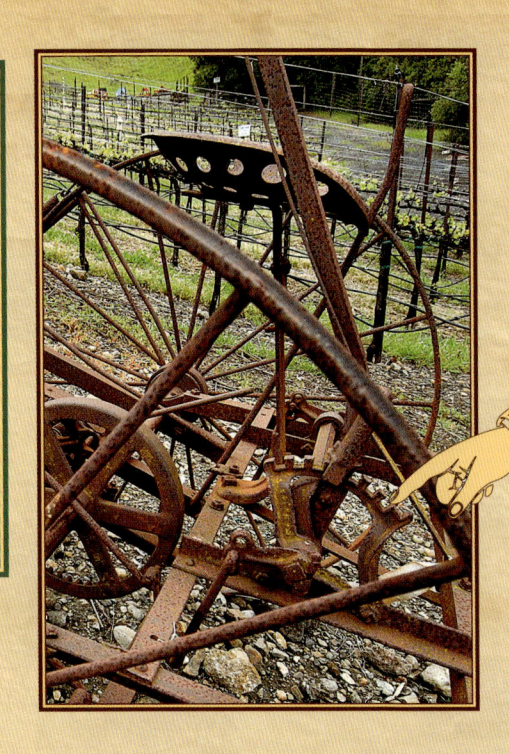

Pull Power Hay Press

John Deere Dain
◄1939►

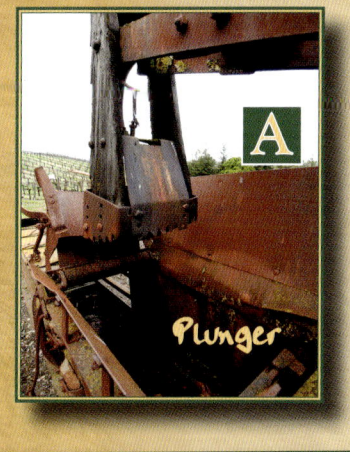

This was a belt-powered, hay-press baler, originally pulled by a team of horses but modifiable for early tractors. Hay was loaded by hand into the plunger chamber and compressed. Then, the hay was pushed through the bale chamber and wood separator blocks were inserted to determine bale length. Once the bales were tied with wire or twine, the blocks were removed.

A. Hay chamber and plunger
B. Bale chamber and wood block.

51

King of the Meadow
McCormick Harvesting Machine Co.
◁1900▷

This hay rake has 32 spring teeth and is 12-feet wide. During the raking process when the hay rake was full, the farmer used the dump handle to leave hay in rows for easy loading, stacking, or hay baling. Almost all hay rakes were called dump rakes.

Hand Dump Rake
Ace Harvester Co.
◁1899▷

Known as a "Laddie Hay Rake," the Hand Dump Rake made straight rows while moving back and forth across a field.

A. Adjustable levers
B. Hay rake teeth

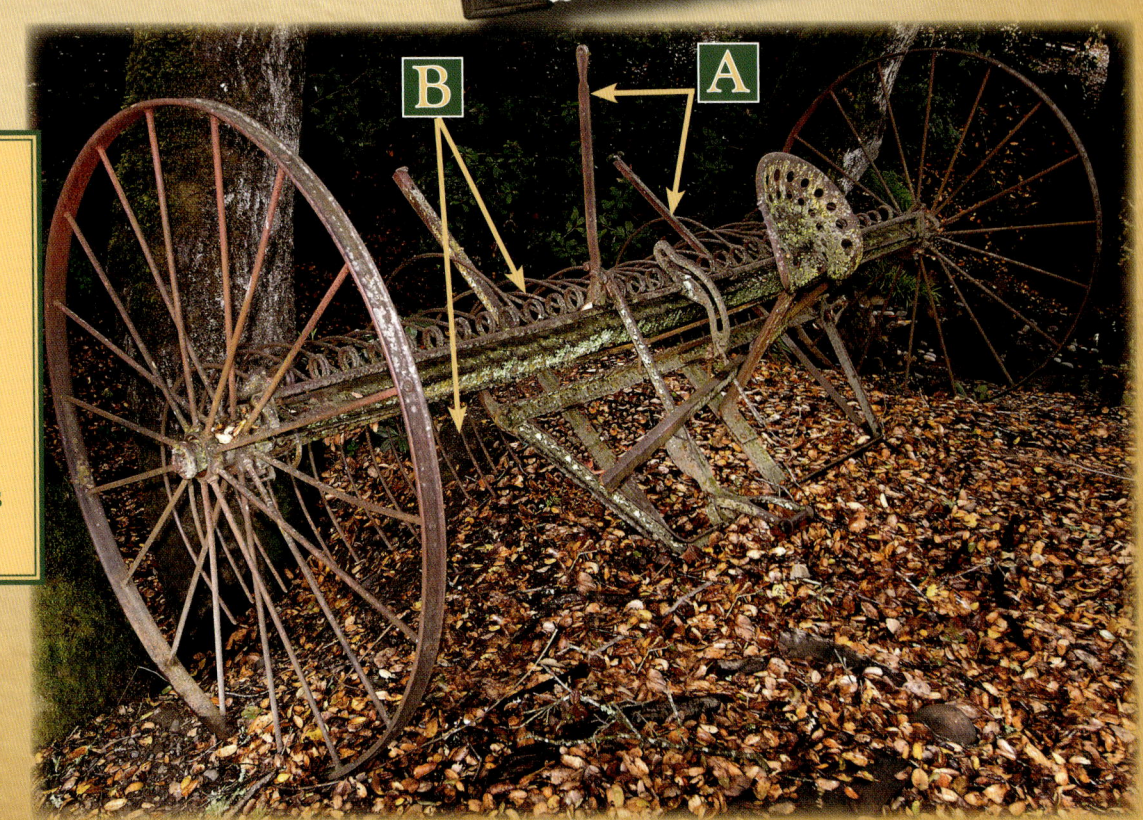

OTHER EQUIPMENT

Subsoil Ripper
Manufacture Unknown
‹1900›

The single ripper shank could cut and tear the subsoil up to a depth of two feet, depending on soil conditions. As many as four horses might be needed to pull the ripper for heavier work. The shank angle was controlled by presetting the dual rotating clickers on each bull wheel.

The purpose of ripping between rows on established vineyards and orchards was to allow water penetration and to create new growth of deeper root systems. Ripping soil was also a common practice prior to any new planting.

A. Bull wheels
B. Ripper shank
C. Rotating clickers

53

Wood Tank Sprayer
Ideal Engine Co.
◁1900▷

Walking alongside the wagon, farmers sprayed orchard trees while directing the horse team that pulled the wagon. The wooden-hull holding tank was powered with a 3½ horsepower gas engine.

A. Gas engine
B. Fill Box

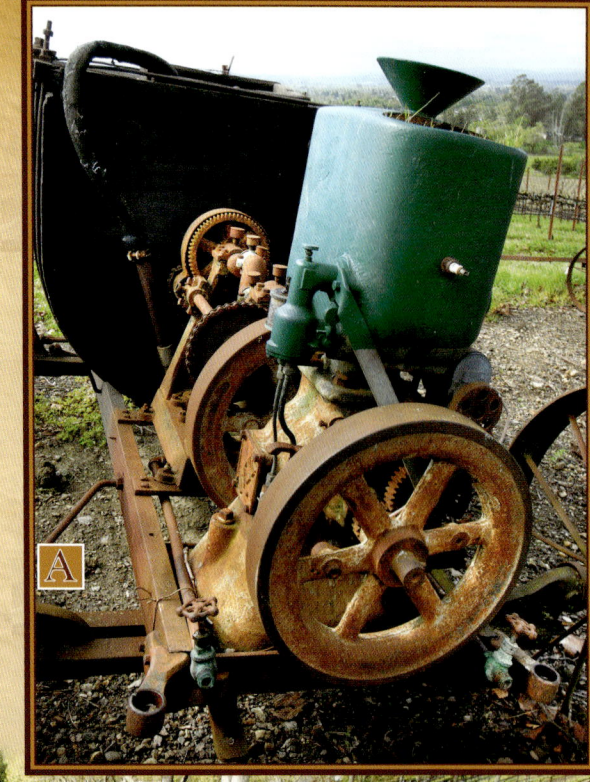

Corn Kernel & Husk Remover
John Deere

This mechanical device removed corn kernels and husks from corn cobs, a job previously done by hand.

Stalk Cutter
Rockwell Co.
1860-1915

The Stalk Cutter was pulled by a team of horses and was used to cut corn stalks after harvest. The vehicle's wheels drove the rotating cyclone cutting blades.

A. Fork fingers in the front were used to align rows of stalks with the cutter. The fingers were spring-loaded to adjust pressure to ground. This also allowed for a smoother ride.
B. The rotating cutting knives could be adjusted by the driver for height to the ground.
C. Control lever for cutting knives.

Windmill
Platter Vale
◄1872►

Using wind energy, this windmill's 18-blade gear drove a shaft up and down into the pump using suction to draw water from the well.

A. Rudder kept the blades turned into the direction of the wind

Pump

Cast iron pumps like this one drew water from suction created by a windmill or a hand pump.

B. Pump shaft hooks to windmill shaft

56

Fertilizer Spreader Model-11
Oliver Farm Equipment
1930

This fertilizer spreader utilized an all-metal 3 x 8-foot box with endless chain conveyor. The conveyor speed was controlled by adjusting the gear ratio setting on the rear wheel axle by using a feed lever. Three rotary beaters broke up and spread the fertilizer.

A. Upper beater
B. Lower beater
C. Main beater
D. Conveyor
E. Conveyor speed control

57

Bemis Plant Setter
Madison Plow Co.
◁1912▷

In days gone by, the driver sat up front on the water barrel with two boys perched in back, placing plants in the furrow. Water was deposited at the plant site through the action of the inner wheel's spaced settings on which a rod rotated, turning the water valve on and off.

A. Seats for two boys placing plants
B. Driver's seat could be removed to allow funnel to be placed in the water barrel *(as shown)*
C. Water drop behind furrow blades with twin curved wheels that shoved dirt over the furrow and compacted dirt over plant roots
D. Water valve and rod

Fordson Tractor
Henry Ford & Son, Inc.
1917 ▷

In 1910 Henry Ford, with his son Edsel, formed a new company, Henry Ford and Son, Inc., to design and mass-produce tractors. In 1917, they introduced the Fordson, the world's first small, lightweight, affordable tractor.

The Fordson tractor used kerosene fuel and had approximately 14 drawbar horsepower. For all its popularity, it had a reputation for hard starting and rearing up and over.

59

Ox-Head Yoke
Local Millwright
◂1800▸

Each side of the yoke was attached to a draft animal—horse, mule or ox—with the center ring attached to the equipment to be pulled.

Single Trees
Local Millwright
Date Unknown

The center section of the single tree attached to the equipment being pulled. The two ends of the single tree hooked up to a horse harness, which in turn hooked up to the pulling animals. Depending on the size of equipment being pulled, multiple single trees for each animal might be required.

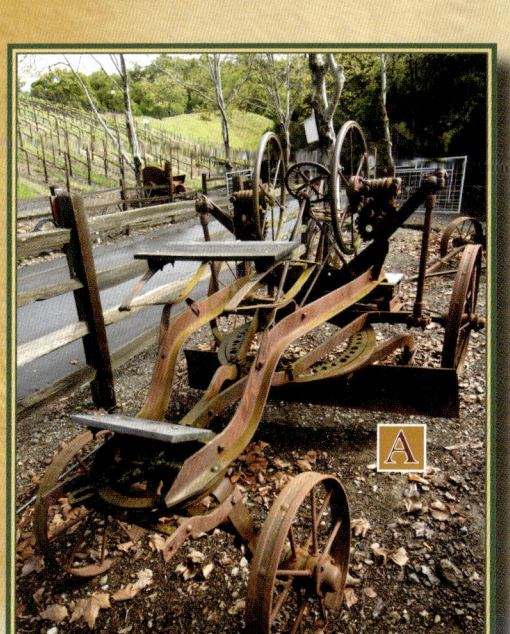

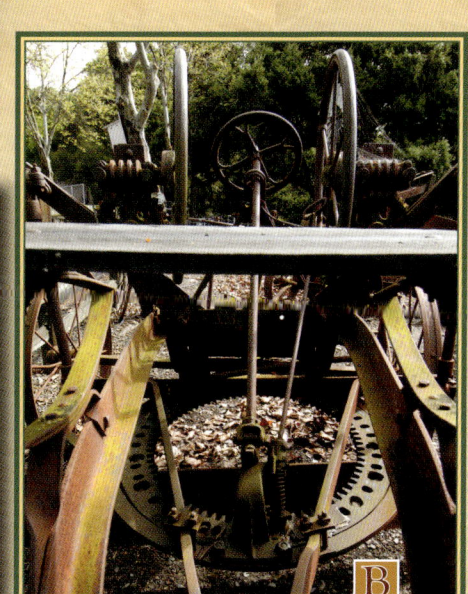

Grader
Western Wheeled Scraper Co.
1900

Drawn by a team of horses, this grader had a four-to-six-foot blade. With an adjustable depth, angle and center pivot, this grader was used in fields and to grade roads. This was the last horse-drawn design before tractors.

A. Blade
B. Blade angle pitch and depth controls
C. Rear axle pivot to blade adjustment
D. Rear wheel axle, left to right adjustment

61

Slip Scraper
P & O Plow Co.
1850-1900

This dirt-moving bucket, commonly known as a slip scraper, was pulled by a two-horse team. The operator raised the handles to set the load.

Manual Fresno
Standard Ditching
◁1892▷

The operator used this implement to push dirt manually to fill ditches or for leveling.

Fresno Canal
Fresno Agricultural Works
◁1891▷

The Fresno Canal was a drag or leveler scraper, normally drawn by one or two horses. The operator used the long leveler to dump his load.

Road Leveler
Western Wheeled Scraper Co.
◁1900▷

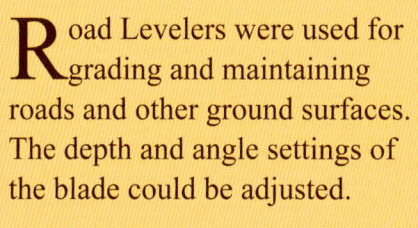

Road Levelers were used for grading and maintaining roads and other ground surfaces. The depth and angle settings of the blade could be adjusted.

A. Operator blade height control
B. Blade height control
C. Blade angle options

63

Flatbed Wagon
Weber Wagon Co.
1845▷

The flatbed wagon, which was used to carry field crops, barrels, and other things, was pulled with two or more horses, depending on the load. The front and rear axles turned separately for sharp cornering. This wagon is shown with a converted tongue for tractor hookup.

A. Dual cornering control rods

Steel Carriage Jack
Peter Henderson & Co.
◁1872▷

This tool was used primarily for wagon wheel repair. The double lift bar was operated by powerful compound levers.

Cargo Wagon
Mitchell Co.
◄1845►

The size of the horse team was determined by the size of the freight load. The Cargo Wagon shown here was a very attractive model with quality hardware. The top cargo rails could be detached so the bottom rails would be shorter for easy loading.

A. Braking system
B. Removable top rails
C. Tailgate for removing dirt, etc.

65

Hand-Drawn Fire Hose Cart
Wirt & Knox
◄1896►

The hand-drawn hose cart transported fire hose from fire to fire. This one has two brass nozzles.

A. Brass nozzle

66

Trapping Implements

Trapping implements like these were used across the nation for hunting fur-bearing animals. The animals' pelts would be made into warm clothing for the family or sold for added income. These were used by Chuck's father to provide additional food and income for his family.

Copper Stills

Copper stills and related accessories were used by Chuck's father and local farmers to make liquor.

The large still shown here has a burner on the bottom with various heads to capture alcohol vapor, depending on the strength desired. Coils attached to the still were made of lead.

The wooden mallet was used for barrel and cork setting.

Also shown is the still's original wooden shipping crate, which was made in Sweden.

Gold Mining Bucket

This was used to haul both ore and men. Two to four men, each standing with one leg inside the bucket and one leg outside, could be transported in and out of the mine shaft. 1849▷

Bucket Winch

The Bucket Winch was used to lower and raise men out of mine shafts. Buckets of dirt and ore were removed the same way. This winch was made and used by Chuck's grandfather in Sonora California. 1930▷

Markers

Markers were used to create lines in the dirt before planting rows of trees, grape vines, and other crops. For equidistant rows, a marker would be mounted on a horse-drawn wagon or other farm equipment and pulled in a straight line.
Millwrights & Blacksmiths

Wine Press

Used to make ciders and wines.

European Wine Press

This is an antique hand-cranked European Wine Press. It was common for one man to produce 150 gallons of grape juice per day for wine.

Scythe & Stable Fork

The Scythe was used for cutting hay grain and grass. The wooden Stable Fork was used as a pitch fork.

Forge

Forges were used by blacksmiths to heat metal objects, such as horseshoes, to achieve a temperature that allowed them to be pounded or bent into desired shapes. This forge unit has pipe legs and a hearth with bellows attached. Champion Blower & Forge Co. ◁1800▷

Timber Saws

For cutting large timber, two men would work together using dual-operator hand-pulled timber saws like these.

Ice Cutter & Ice Hook

In the days before modern refrigeration, nature's ice provided necessary cooling. Taking advantage of frozen waterways, ice cutters were used to harvest blocks of ice, which were lifted by ice hooks. ◁1900▷

Automatic Clothes Washer & Boiler

This is a manually operated laundry tub, advertised as "automatic" because it was a step up from the washboard. 1868 ▷

Wheel Barrow Sprayer

Using this manual hand pump (now missing its hose) grapevines were sprayed for fungus by pumping the handle up and down to suction liquid into the hose. F.E. Meyers & Brother ◁1900▷

Hand pump

69

Index

Automatic Clothes Washer & Boiler	69	European Wine Press	68
Bemis Plant Setter	58	Fertilizer Spreader - Model 11	57
Bucket Winch	68	Flatbed Wagon	64
Cargo Wagon	65	Fordson Tractor	59
Check Row Marker	14	Forge	69
Copper Stills	67	Fresno Canal	62
Corn Kernel & Husk Remover; Stalk Cutter	55	Gold Mining Bucket	68
		Grader	61
Cultivator		Hand-Drawn Fire Hose Cart	66
Belanger	24	Harrow	
Belanger Steel Scuffler	24	A-Frame	27
Buckeye, American Seeding	21	Cultivator	32
Complex	23	Iron Age	31
Double Fork Rake	19	Pennsylvania Disc	29
Dual Purpose, Brown Man	15	Revolving	27
Garden, Case-Keeler	17	Spike Tooth	32
Garden, Henderson	17	Spring Tooth	26
Jack Rabbit, Avery	20	Sulky	30
Planet Junior, No. 101, Allen	16	Ten Blade Disc	28
Planet Junior, No. 101, Henderson	16	Haying Equipment	
Planet Junior No. 101 Horse Hoe	25	California Grapple Fork	45
Planter's Pride, Ohio Cultivator	18	Hand Dump Rack	52
Small, Allen	22	Hay Blower	47
Surface, Double Fork Rake, Deere	19	Hay Rake/Sling	46
		Hay Tedder	50
		Ideal Mower	49
		King of the Meadow	52
		Pull Power Hay Press	51
		Sandwich Hay Mower	48

Chuck fishing on the ranch, c. 1940.

Index

Ice Cutter and Ice Hook	69	Walking Furrow	34
Manual Fresno	62	Wooden Double Shovel	34
Markers	68	Road Leveler	63
Ox-Head Yoke	60	Scythe & Stable Fork	68
Plow		Seed Planter	
Cambridge Reversible	43	Avery Sure Drop Planter	11
Canton Clipper	35	Grain or Seed Drill	12
Ditch Cleaner	38	Iron Ace	10
Ditch Irrigation	38	Keystone Corn Planter	13
Double Mold	37	Pratt's Seed Sower	6
Furrow	36	Seed Planter, Oliver	9
Improved Winged Shovel &		Single Row Seeder, Henderson	7
Potato Digger Plow	41	Single Seeder, Case Keeler	8
Large Walking	40	Single Trees	60
Oliver Sulky	42	Slip Scraper	62
Potato Digger	39	Stalk Cutter	55
Potato Digger, Walking	39	Steel Carriage Jack	64
Prairie Breaker	33	Subsoil Ripper	53
Red Jacket Walking	36	Timber Saws	69
Reversible	35, 43	Trapping Implements	67
Stirring	37	Wheel Barrow Sprayer	69
Tractor, Gang	44	Windmill & Pump	56
Turning	40	Wine Press	68
Vulcan Chilled	41	Wood Tank Sprayer	54

The Elsbree Ranch branding iron, the "Bell E".

Antique Farm Equipment: The Elsbree Collection is available at bookstores and from the publisher. You can photocopy the order form below and mail it, along with your check, money order, or credit card information, to Rayve Productions; or if you are using a credit card, you can call Rayve Productions toll-free at 800-852-4890, fax your order to 707-838-2220, or order at our website www.rayvepro.com

Price:	$27.95
Shipping:	Priority Mail - $5.50 for the first book + $2.00 each additional book
	Media Mail - $3.99 for the first book + $2.00 each additional book
Sales Tax:	California residents only, please add 8% sales tax.

Rayve Productions   800-852-4890;
local 707-838-6200
fax 707-838-2220
P.O. Box 726, Windsor CA 95492

Order

	Quantity	Amount
Antique Farm Equipment: The Elsbree Collection	_____	$_____
Shipping		$_____
Subtotal		$_____
Sales Tax (Calif. residents only)		$_____
Total		$_____

Name _____ Phone _____

Address _____ Email _____

City, State, Zip _____

____ Check enclosed $_____ Date _____

____ Charge my Visa/MasterCard $_____

Credit Card # _____ Expiration _____

Signature _____

Thank you!